# DOHERTY POWER AMPLIFIERS

# DOHERTY POWER AMPLIFIERS

## From Fundamentals to Advanced Design Methods

BUMMAN KIM

ELSEVIER

ACADEMIC PRESS

An imprint of Elsevier

Academic Press is an imprint of Elsevier
125 London Wall, London EC2Y 5AS, United Kingdom
525 B Street, Suite 1800, San Diego, CA 92101-4495, United States
50 Hampshire Street, 5th Floor, Cambridge, MA 02139, United States
The Boulevard, Langford Lane, Kidlington, Oxford OX5 1GB, United Kingdom



**Library of Congress Cataloging-in-Publication Data**
A catalog record for this book is available from the Library of Congress

**British Library Cataloguing-in-Publication Data**
A catalogue record for this book is available from the British Library

ISBN: 978-0-12-809867-7



*Publisher:* Mara Conner
*Acquisition Editor:* Tim Pitts
*Editorial Project Manager:* Leticia Lima
*Production Project Manager:* Surya Narayanan Jayachandran
*Cover Designer:* Matthew Limbert

Typeset by SPi Global, India

# CONTENTS

# ACKNOWLEDGMENTS

Many persons deserve warm thanks for making this book a reality. The major part of this book contains the results of more than a decade of research activities performed in my group at POSTECH. I have been fortunate from the support of numerous excellent students, who devote their time and energy to work on Doherty power amplifiers. I would like to express my sincere gratitude to all those people who have worked on Doherty power amplifier. The deliberate work is the basis of this book.

I would like to express my sincere appreciation to all the Academic Press staff involved in this project for their cheerful professionalism and outstanding efforts. Last but not the least, I would like to thank my very nearest family for their patience and understanding during the many days spent working on this book.

CHAPTER ONE

# Introduction to Doherty Power Amplifier

## 1.1 HISTORICAL SURVEY

William H. Doherty was an American electrical engineer, best known for his invention of the Doherty amplifier. Doherty was born in Cambridge, Massachusetts, in 1907. He attended Harvard University, where he received his bachelor's degree in communication engineering (1927) and his Master's degree in engineering (1928). Doherty joined Bell Labs in 1929. At the Bell Labs., he worked on the development of high-power radio transmitters, which were used for transoceanic radio telephones and broadcastings.

Doherty invented this unique amplifier approach in 1936 using a vacuum tube amplifier. This new device greatly improved the efficiency of RF power amplifiers and was first used in a 50 kW transmitter that Western Electric Company designed for WHAS, a radio station in Louisville, Kentucky. Western Electric went on to incorporate Doherty amplifiers into, at least, 35 commercial radio stations worldwide by 1940 and many other stations, particularly in Europe and Middle East in the 1950s. Within the Western Electric, the device was operated as a linear amplifier with a driver that was modulated. In the 50 kW implementation, the driver was a complete 5 kW transmitter that could, if necessary, be operated independently of the Doherty amplifier, and the Doherty amplifier was used to raise the 5 kW level to the required 50 kW level.

As a successor to Western Electric Company Inc. for radio broadcast transmitters, the Continental Electronics Manufacturing Company at Dallas, Texas, considerably refined the Doherty concept. The early Continental Electronics designs, by Weldon and others, retained most of the characteristics of Doherty's amplifier but added medium-level screen-grid modulation of the driver. The ultimate refinement made by the company was the high-level screen-grid modulation scheme invented by Sainton, whose transmitter consisted of a class C carrier tube in a parallel connection with a class C peaking tube. The tubes' source (driver) and load (antenna) were split and combined through + and $-90°$ phase-shifting networks as in a Doherty amplifier. The unmodulated radio-frequency carrier was applied to the control grids of both tubes. Carrier modulation was applied to the screen grids of both tubes, but the screen-grid bias points of the carrier and peaking tubes were different and were established such that the peaking tube was cut

*Doherty Power Amplifiers*
https://doi.org/10.1016/B978-0-12-809867-7.00001-6

off when modulation was absent and the amplifier was producing rated unmodulated carrier power. And both tubes were conducting during the modulation. And each tube was contributing twice the rated carrier power during 100% modulation as four times the rated carrier power is required to achieve 100% modulation. As both tubes were operated in class C, a significant improvement in efficiency was thereby achieved in the final stage. In addition, as the tetrode carrier and peaking tubes required very little drive power, a significant improvement in efficiency within the driver was achieved as well. The commercial version of the Sainton amplifier employed a cathode-follower modulator, and the entire 50 kW transmitter was implemented using only nine total tubes of four tube types, a remarkable achievement, given that the transmitter's most significant competitor from RCA was implemented using 32 total tubes of nine tube types.

The approach was used by such leading companies as not only Continental but also Marconi with functional installations up to the late 1970s. The IRE recognized Doherty's important contribution to the development of more efficient radio-frequency power amplifiers with the 1937 Morris N. Liebmann Memorial Award.

The amplifier has been reinvented recently for use in mobile communication systems using semiconductor devices at higher frequencies. It creates large deviations from the previous design based on the vacuum tubes. Also, the amplifier is modified to amplify a highly modulated signal with a high peak-to-average power ratio (PAPR). Nowadays, the Doherty amplifier is the choice of the technique for the power amplification in the mobile base-station. The technology can be useful for handset power amplifier, also. In this chapter, the basic structure of the Doherty amplifier together with the operational behavior is introduced.

## 1.2 BASIC OPERATION PRINCIPLE

The most important property of the Doherty amplifier is the load modulation, which carries out the perfect combining of the asymmetrical powers from the two amplifiers. Thereby, only one amplifier (called carrier amplifier) operates at a low power level, and the efficiency at the same power level is two times higher than that obtained from the two times bigger amplifier. The two amplifiers (the second one is called peaking amplifier) generate powers at a higher power level, and the carrier amplifier operates at the peak efficiency mode in this region due to the nice load modulation characteristic. This property provides an efficient amplification of an amplitude-modulated signal. The load, which is modulated by the current ratio of the carrier and peaking amplifiers, is self-adjusted for the peak efficiency at the two power levels. The first peak efficiency is provided by the carrier amplifier (CA) at the level when the peaking amplifier (PA) is turned on, and the second peak is at the power level when the two amplifiers generate their full powers. Another important characteristic of the Doherty load modulation is that the overall gain of the amplifier is constant, providing a linear amplification.

## 1.2.1 Load Modulation Behavior

### *1.2.1.1 Load Impedance Modulation*

The simplest illustration of the load modulation concept is shown in Fig. 1.1, where a voltage-controlled voltage source (VCVS) is in parallel with a voltage-controlled current source (VCCS) and a load resistor $R$. The impedance seen by the VCVS, $Z_1$, is modulated by the current $I_2$, as given by

$$Z_1 = \frac{V_1}{I_1} = \frac{V_1}{I_R - I_2} \tag{1.1}$$

Varying the current $I_2$ from zero to $I_R = V_1/R$, $Z_1$ is varied from $R$ to $\infty$. In this circuit, the VCCS modulates the load impedance of the VCVS. In Doherty amplifier, the ability to modulate $Z_1$ using $I_2$ is properly employed to track the optimal impedances for the amplifier to operate efficiently at the back-off power levels. An important property of the setup in Fig. 1.1 is that the linearity of the overall system is solely determined by the linearity of the VCVS because the voltage $V_{out}$ across the load is always equal to $V_1$. Therefore, linearity is guaranteed regardless of the value of $I_2$, as long as $V_1$ is linearly proportional to $V_{in}$. For this purpose, the impedance $Z_1$ should track a given impedance profile versus $V_{in}$ by specifying the $I_2$ versus $V_{in}$ profile. Although mathematically simple to define it, realizing a given $I_2$ versus $V_{in}$ profile in practice can be a challenge.

In the load modulation technique, the VCVS and VCCS have their important roles. The former ensures the linearity of the amplifier, while the latter acts as the load modulating device, whose $I_2$ versus $V_{in}$ profile determines the impedance $Z_1$ seen by the VCVS. These two properties are important in derivation of the Doherty circuit configuration.

The Doherty amplifier uses a different circuit topology for the load modulation. It consists of two amplifiers (two current sources) and an impedance-inverting network, which converts the one current source to a voltage source. This converted amplifier is called a carrier amplifier, and the other current source amplifier is a peaking amplifier.

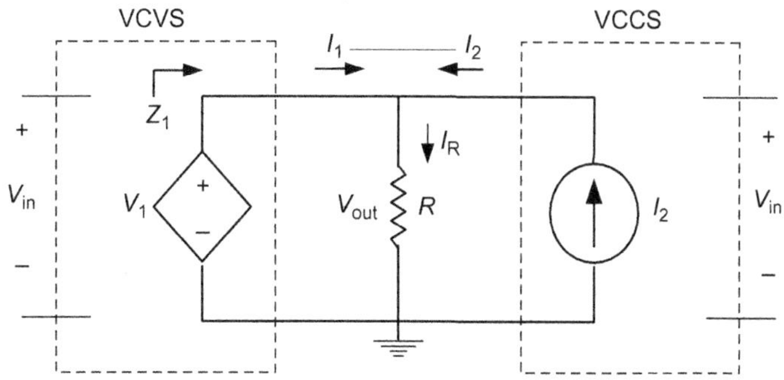


**Fig. 1.1** Load modulation circuit driven by voltage and current sources.

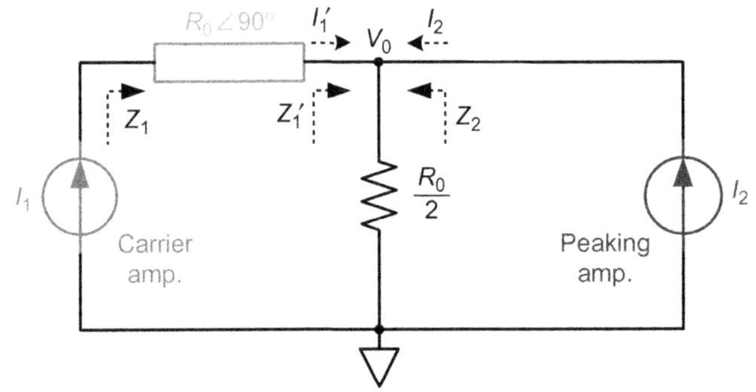


**Fig. 1.2** Operational diagram of Doherty amplifier.

Fig. 1.2 shows an operational diagram to analyze the Doherty amplifier circuit. The output load is connected to the carrier amplifier through the impedance inverter (a quarter-wave transmission line) and directly to the peaking amplifier. In this figure, the optimum power-matching impedance of the carrier and peaking amplifiers at the peak power is $R_0$, and the load of the carrier amplifier, when the peaking amplifier is off, becomes $R_0/2$ due to the parallel connection of the two amplifiers. It is assumed that the output capacitor of the device is resonated out and the phase delay of the quarter-wave line is compensated at the input.

The impedance inverter has a characteristic impedance of $R_0$ also. The load impedances of the carrier amplifier at $Z_1'$ and $Z_1$, shown in Fig. 1.2, are given by

$$Z_1' = \frac{V_0}{I_1'} = \frac{R_0}{2} \cdot \left[\frac{I_1' + I_2}{I_1'}\right] \tag{1.2}$$

$$Z_1 = \frac{{R_0}^2}{Z_1'} = \frac{2R_0}{\left(1 + \dfrac{I_2}{I_1'}\right)} = \frac{2R_0}{(1+\alpha)} \tag{1.3}$$

where $\alpha = {}^{I_2}/_{I_1'}$. Eq. (1.3) shows that the carrier amplifier represented by a current source $I_1$ sees the load impedance modulated by the second current source $I_2$, representing the peaking amplifier. It should be noticed that $I_1'$ is different from $I_1$ due to the impedance change. Also, in the normal Doherty operation, the current level of the peaking amplifier varies from 0 to $I_1 = I_{max}$, the maximum current of the two amplifiers, and $\alpha$ changes from 0 to 1. Normally, $I_1$ and $I_2$ can handle the same amount of current, that is, the same size devices for the two amplifiers, and $Z_1$ is $R_0$ when $I_2 = I_1 = I_{max}$ at the peak power, because $I_1$ is equal to $I_1'$ at the power. $Z_1$ is $2R_0$ when $I_2 = 0$ and $Z_1$ is in between the two values for the $I_2$ current between 0 and $I_{max}$. This is the Doherty load modulation behavior, which is depicted in Fig. 1.3C.

The peaking amplifier provides the open load until it is turned on because the current $I_2$ is zero. After turned on, the impedance $Z_2$ is also modulated similarly, which is given by

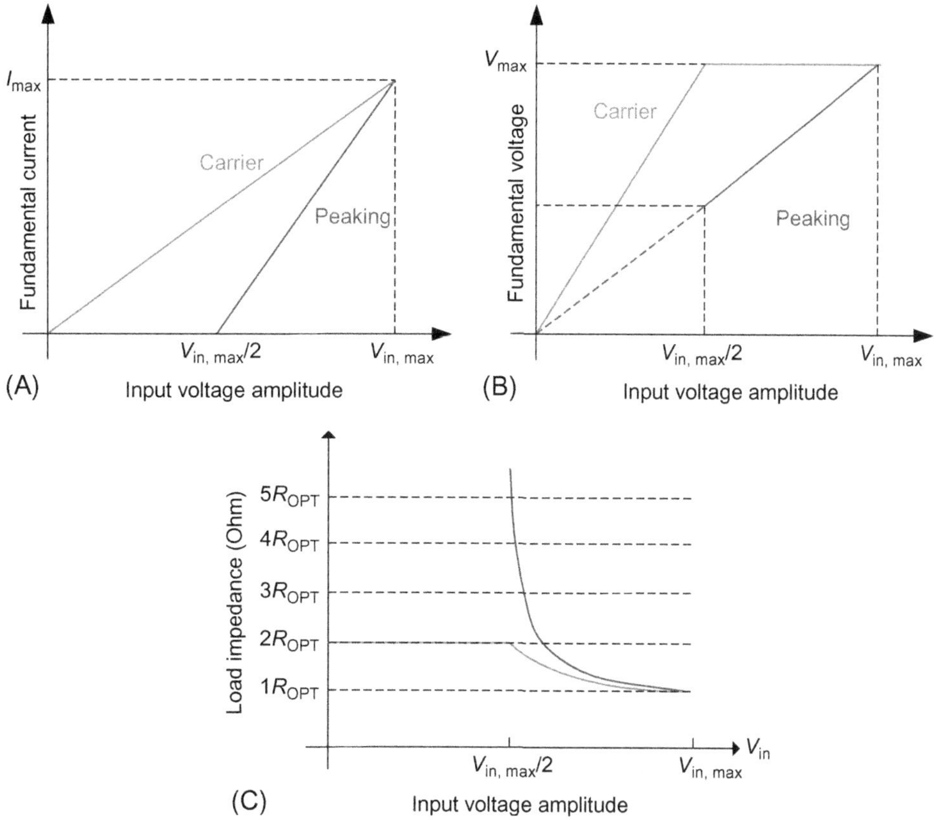


**Fig. 1.3** Current, voltage, and load impedance shapes of the carrier and peaking amplifiers: (A) current profiles, (B) voltage profiles, (C) load impedance profiles.

$$Z_2 = \frac{V_0}{I_2} = \frac{R_0}{2} \cdot \frac{(I_1' + I_2)}{I_2} = \frac{R_0}{2} \cdot \frac{(1 + \alpha)}{\alpha} \tag{1.4}$$

The load modulation behavior is also depicted in Fig. 1.3C. The carrier impedance is modulated from $2R_0$ to $R_0$ and the peaking from infinity to $R_0$. In this figure, it is assumed that each current source is linearly proportional to the input voltage and $R_0$ is equal to $R_{OPT}$ of the transistor, the optimum power matching resistance. As shown in Fig. 1.3A, $I_2$ is turned on at the midpoint due to the class C bias of the peaking amplifier and is increased to the maximum value. $I_1$ is linearly increased from the zero gate voltage due to the class B bias. In this operation, the transconductance of the peaking amplifier should be twice larger than that of the carrier amplifier due to a half of the input voltage swing for the maximum current generation. To get the two times larger transconductance, the peaking amplifier should be two times larger than the carrier amplifier. But in the case, only a half of the peaking current is utilized, wasting the power generation capability. To solve the problem, uneven driving technique is developed, which will be introduced in Chapter 2.

### *1.2.1.2 Voltage, Current, and Load Impedance Profiles*

Under the load modulation described in Section 1.2.1.1, the current and voltage profiles of the carrier and peaking amplifiers are depicted in Fig. 1.3. Fig. 1.3A shows the fundamental currents of the carrier and peaking amplifiers, $I_C$ and $I_P$, respectively. The two devices generate the same maximum current as shown in the figure. The currents are given by

$$I_1 = I_{\max}\left[\frac{V_{\text{in}}}{V_{\text{in,max}}}\right] \quad 0 < V_{\text{in}} < V_{\text{in,max}} \tag{1.5}$$

$$I_2 = \begin{cases} 0 & 0 < V_{\text{in}} < V_{\text{in,max}}/2 \\ I_{\max}\left[\dfrac{V_{\text{in}} - V_{\text{in,max}}/2}{V_{\text{in,max}}/2}\right] & V_{\text{in,max}}/2 < V_{\text{in}} < V_{\text{in,max}} \end{cases} \tag{1.6}$$

By introducing the current profiles to Eqs. (1.3), (1.4), the load impedances of the carrier and peaking amplifiers, $Z_1$ and $Z_2$, can be calculated:

$$Z_1 = \begin{cases} 2R_0 & 0 < V_{\text{in}} < V_{\text{in,max}}/2 \\ \dfrac{2R_0}{1+\alpha} & V_{\text{in,max}}/2 < V_{\text{in}} < V_{\text{in,max}} \end{cases} \tag{1.7}$$

$$Z_2 = \begin{cases} \infty & 0 < V_{\text{in}} < V_{\text{in,max}}/2 \\ \dfrac{R_0}{2}\cdot\dfrac{(1+\alpha)}{\alpha} & V_{\text{in,max}}/2 < V_{\text{in}} < V_{\text{in,max}} \end{cases} \tag{1.8}$$

$$\text{with } \alpha = {}^{I_2}\!/_{I_1'} = \frac{2R_0}{Z_1}\left[\frac{V_{\text{in}} - V_{\text{in,max}}/2}{V_{\text{in}}}\right] \tag{1.9}$$

The voltage profiles of the carrier and peaking amplifiers can be calculated using the current profiles and load impedances of $Z_1$ and $Z_2$, respectively, which are given by

$$V_1 = \begin{cases} 2R_0\cdot\left(\dfrac{I_{\max}}{V_{\text{in,max}}}\right)\cdot V_{\text{in}} & 0 < V_{\text{in}} < V_{\text{in,max}}/2 \\ R_0\cdot I_{\max} & V_{\text{in,max}}/2 < V_{\text{in}} < V_{\text{in,max}} \end{cases} \tag{1.10}$$

$$V_2 = \left\{ R_0\cdot\left(\frac{I_{\max}}{V_{\text{in,max}}}\right)\cdot V_{\text{in}} \quad 0 < V_{\text{in}} < V_{\text{in,max}} \right. \tag{1.11}$$

Because the load impedance of the carrier amplifier is $2R_0$ at a low-power operation, the voltage approaches the peak level when the peaking is turned on. After that, the voltage remains at the peak level because the load resistance decreases accordingly as the current increases. The voltage across the peaking amplifier increases linearly and reaches to the peak level when the input voltage is at the maximum. In $0 < V_{\text{in}} < \frac{V_{\text{in,max}}}{2}$, the voltage of the peaking amplifier is not defined because the peaking amplifier is turned off. But the voltage should be $V_0$, which is converted from $V_1$ in Eq. (1.10). Overall, the voltage profile is given by Eq. (1.11), which is determined by the current of the input voltage multiplied by $2 \cdot G_m$, the effective transconductance, and the load impedance of $R_0/2$.

#### *1.2.1.3 Load Lines for the Modulated Loads*

The load lines of the two amplifiers can be calculated from the voltage and current profiles shown in Fig. 1.3 and are presented in Fig. 1.4. Due the load modulation, the load line for the carrier amplifier follows $2R_0$ until the peaking amplifier is turned on. After the peaking is turned on, the load of the carrier is reduced following Eq. (1.7). Due to the reduced load impedance according to the current level, the load line follows the knee region as shown in the figure, maintaining the high efficiency of the class B amplifier. The peaking amplifier is off at the low power with the open load. As soon as it is turned on, the voltage is built on fast with the transconductance of 2Gm and load impedance of Ro/2 and reaches to the peak voltage at the peak input voltage following Eq. (1.11), which is shown in Fig. 1.4.

### 1.2.2 Efficiency and Gain Characteristics

As shown in Fig. 1.2, the Doherty load modulation circuit perfectly combines the asymmetrical output powers from the carrier and peaking amplifiers in a current-combining way. This asymmetrical power-combining capability, which is not easily realized in other ways, is a big merit of the Doherty modulation circuit. Thus, the carrier amplifier operates with a high efficiency at a low-power region and the maximum efficiency at the higher-power region. Moreover, this load modulation provides a constant gain for a linear amplification.

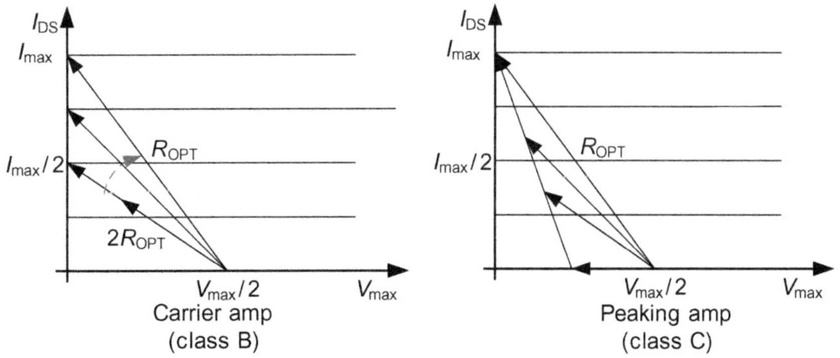


**Fig. 1.4** Load-lines of the carrier and peaking amplifiers according to the input voltage.

### *1.2.2.1 Efficiency*

Generally, a power amplifier, with a specific optimum load impedance, delivers the maximum efficiency only at the peak power since the full output voltage and current swings are achieved only at the peak output power defined by the fixed load. On the other hand, the load of the Doherty power amplifier is dynamically modulated, providing the peak efficiency at the different power level determined by the modulated load. For the Doherty load modulation, the carrier amplifier provides the maximum efficiency at a half of the maximum input voltage, and the maximum efficiency is maintained through the load modulation at a higher power as shown in Fig. 1.5. The peaking amplifier achieves the maximum efficiency only at the peak power.

In the low-power region ($0 < V_{in} < V_{in,\ max}/2$), the peaking amplifier remains in the cutoff state, and the load impedance of the carrier amplifier is two times larger than that of the conventional amplifier. Thus, the carrier amplifier reaches to the saturation state at the input voltage of $V_{in,\ max}/2$ since the maximum voltage swing reaches to $V_{dc}$ with a half of the maximum current level. As a result, the maximum power level is a half of the carrier amplifier's allowable power level (a quarter of the total maximum power or 6 dB down from the total maximum power), and the efficiency of the Doherty amplifier is equal to the maximum efficiency of the class B carrier amplifier.

In this low-power operation, only the carrier amplifier with a half of the total power cell produces power, and the efficiency at a given power level is twice higher than that of the two times bigger power cell amplifier.

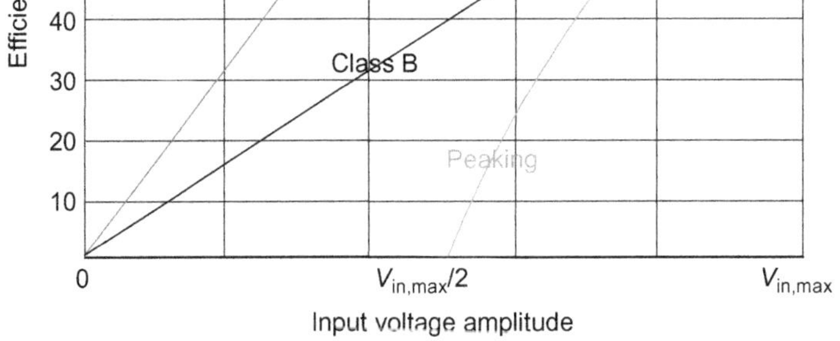

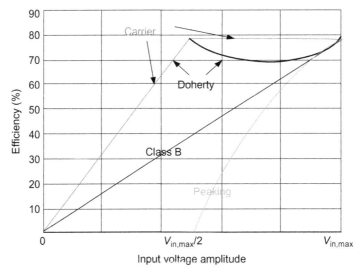

**Fig. 1.5** Plot of efficiencies versus input drive level for the Doherty amplifier with the class-B-biased carrier amplifier, the class-C-biased peaking amplifier and the class B amplifier.

In the higher-power region ($V_{in,\ max}/2 < V_{in} < V_{in,\ max}$), where the peaking amplifier is conducting, the carrier amplifier generates power with the peak efficiency since the saturated operation is maintained due to the load modulation. The peaking amplifier operates with very large load at the turn-on stage, and the efficiency increases very fast but is still lower than the peak efficiency. Thus, the total efficiency of the Doherty amplifier is degraded. But the second maximum efficiency point is reached when the peaking amplifier provides the maximum efficiency at the peak power with saturated operation. Therefore, it has two maximum efficiency points, enhancing the efficiency at the backed-off output power level as shown in Fig. 1.5. The efficiency of the Doherty amplifier at the maximum input voltage is equal to the maximum efficiency of the conventional amplifiers. Therefore, the Doherty amplifier provides higher efficiency over the whole power ranges compared with a conventional class B amplifier. The resulting Doherty amplifier can solve the problem of low efficiency for amplification of a large PAPR signal.

#### 1.2.2.2 Gain

To drive the carrier and peaking amplifiers with an equal power, the input power of the Doherty amplifier is divided using a 3 dB coupler. Therefore, the input power level to the carrier is 3 dB lower than the total input power. However, the load of the carrier is two times higher than $R_{OPT}$ at a low-power operation, and the gain is increased by 3 dB by the $2R_{OPT}$ load, recovering the gain loss due to the 3 dB lower input power. At the peak-power operation, the carrier and peaking amplifiers become a two-way current power-combining structure with $R_{OPT}$ load, and the gain is the same as at a low power level. In between the two power level, due to the nice asymmetrical power combining with the load modulation, the gain is maintained at the constant value.

The power gain at the intermediate-drive region can be calculated from Fig. 1.2:

$$I_1' = I_1 \cdot \sqrt{\frac{Z_1}{Z_1'}} = I_1 \frac{2}{1+\alpha} \tag{1.12}$$

The current flowing through the load $R_0/2$, $I_L$, is given by

$$I_L = I_1' + I_2 = I_1' + \alpha I_1' = 2 \cdot I_1 \tag{1.13}$$

The $2 \cdot I_1$ current flows through $R_0/2$ load during the load modulation. Under the condition that the input voltage generates linear currents $I_1$ as shown in Fig. 1.3A, the constant gain is maintained during the load modulation. Due to the constant gain characteristic, the ideal Doherty amplifier is a linear amplifier, if the nonlinearity of the transistor is not considered.

# 1.3 OFFSET LINE TECHNIQUE

## 1.3.1 Realization of Doherty Amplifier

So far, it is assumed that a transistor is a current source with the optimum power-matching impedance of $R_{OPT}$ and the load impedance $R_L$ of the Doherty amplifier is *equal to* $R_{OPT}/2$. To realize a Doherty amplifier using a transistor, the inputs and outputs of the carrier and peaking amplifiers are matched to $R_0 = 50\ \Omega$ load at the peak power, and the Doherty load modulation circuit is attached after the matching circuits as shown in Fig. 1.6. The two parallel-connected $R_0$ impedances are transformed to $R_0/2$ and the impedance is transferred to $R_0$ by an impedance inverter ($R_T = R_0/\sqrt{2}$). Therefore, the Doherty amplifier is matched properly at the peak-power operation. During the load modulation of the Doherty network, the impedance at the carrier amplifier after the matching circuit is modulated from $R_0$ to $2R_0$. For the modulated load impedance of $R_0{\sim}2R_0$, the impedance should be transferred to $R_{OPT}{\sim}2R_{OPT}$ at the drain by the matching circuit, but the impedance is not transferred in that way. The load modulation of the peaking amplifier is also deviated from the ideal modulation of $R_{OPT}{\sim}\infty$. It is known that the transfer is properly carried out only for the real impedance system without imaginary part. To handle the complex impedance system, the offset line technique is introduced.

To produce the real impedance system, the output capacitance of a device is resonated at the drain terminal using a parallel inductor and the Doherty network, and the matching circuit is attached, thereafter. This approach is popularly employed in on-chip Doherty amplifier. But the Doherty amplifier for the base-station application using packaged device cannot employ that technique, and the offset line should be employed.

Fig. 1.6 shows a schematic of the Doherty amplifier with the input- and output-matching circuits. The input power is divided to the carrier and peaking amplifiers using

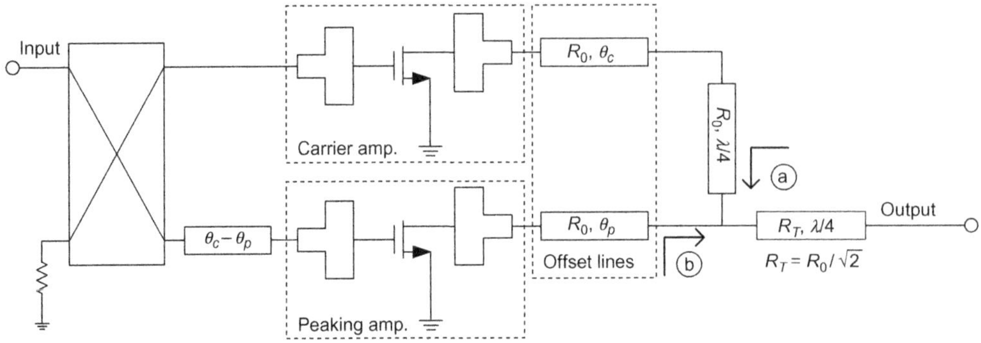


**Fig. 1.6** Schematic diagram of the Doherty amplifier.

a 3 dB coupler. The inputs of the two amplifiers are matched to the coupler with 50 Ω like a conventional two-way power-combined amplifier. The two outputs are matching to 50 Ω, and the Doherty network is attached. To provide the proper load modulation, the offset lines are attached between the matching circuits and the Doherty modulation circuit. The $R_0/2$ is transferred to $R_0$ using an inverter $R_T$. The phase mismatch between the two paths is corrected at the input.

### 1.3.2 Operation of the Offset Line

The offset lines are microstrip lines having a characteristic impedance equal to the load impedance ($R_0$) of the carrier or peaking amplifiers at the peak-power operation with a proper length. While the load impedance is modulated, the line rotates the impedance plane of the matching circuit to the real axis. This rotation eliminates the phase produced by the complex impedance match of the device, realizing the proper load modulation. The line produces a phase-delay difference between the carrier and peaking amplifiers, but the delay can be easily adjusted at the input.

#### *1.3.2.1 Offset Line at Carrier Amplifier*

The offset lines for the carrier amplifier and peaking amplifiers are shown in Fig. 1.7. The functions of the two lines are identical. Let's consider the output section of the carrier amplifier. The matching network compensates the device reactive and parasitic elements and transforms a load $R_C = R_0$ to the desired $Z_{IN,c} = R_{OPT}$ at the current source. However, for the load of $R_0$ to $2R_0$, $Z_{IN,c}$ is not converted to the resistive impedance but becomes a complex impedance. Without the offset line, the input reflection coefficient of the scattering matrix $S$ of the cascade of the device reactive part and matching network (Fig. 1.7B) is given by

$$\Gamma_{IN,c} \triangleq \frac{Z_{IN,c} - R_{OPT}}{Z_{IN,c} + R_{OPT}} = \frac{S_{11} - \Delta_S \Gamma_C}{1 - S_{22}\Gamma_C} \tag{1.14}$$

where $\Delta S$ is the determinant of the $S$-parameter with a characteristic impedance of $R_{OPT}$ and $\Gamma_C$ is the reflection coefficient of $R_C$ with respect to $R_{OPT}$, which is always a real number. The matching network is designed to ensure $\Gamma_{IN,c} = 0$ when $\Gamma_C = \Gamma_0$ (load of $R_0$). Since the network represented by $S$ can be considered as purely reactive, we have $S_{22} = \Gamma_0$. The lossless nature of the network implies also that $S$ is Hermitian; thus,

$$S_{11}S_{21}^* + S_{12}S_{22}^* = 0 \quad \text{and} \quad |S_{22}|^2 + |S_{21}|^2 = 1 \tag{1.15}$$

Since the network is reciprocal ($S_{12} = S_{21}$) and $S^*_{22} = S_{22}$, Eq. (1.15) leads to

$$S_{21}^2 = -\frac{S_{11}S_{21}^*}{S_{22}} S_{21} \Rightarrow \frac{S_{11}}{S_{22}} = -\frac{S_{21}^2}{|S_{21}|^2} = -e^{j2\angle S_{21}} \tag{1.16}$$

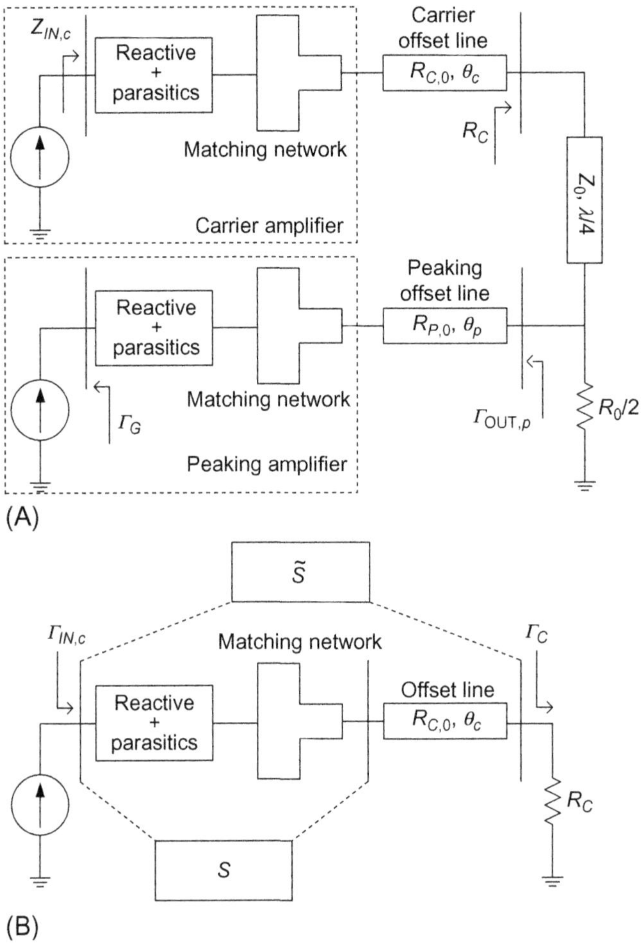


**Fig. 1.7** (A) Output section of the Doherty amplifier. (B) Output section of the carrier amplifier.

Using Eqs. (1.15), (1.16), the determinant $S$ can be represented as

$$\Delta S = S_{11}S_{22} - S_{21}^2 = \frac{S_{11}}{S_{22}}\left(S_{22}^2 + |S_{21}|^2\right) = \frac{S_{11}}{S_{22}} \tag{1.17}$$

Eq. (1.14) can be rearranged as

$$\Gamma_{IN,c} = \frac{S_{11}}{S_{22}}\frac{S_{22} - \Gamma_C}{1 - S_{22}\Gamma_C} = \frac{\Gamma_C - S_{22}}{1 - S_{22}\Gamma_C}e^{j2\angle S_{21}} \tag{1.18}$$

Note that

$$\frac{\Gamma_C - S_{22}}{1 - S_{22}\Gamma_C} = \frac{\Gamma_C - \Gamma_0}{1 - \Gamma_0\Gamma_C} = \frac{R_C - R_0}{R_C + R_0} = \Gamma_{C|R_0} \tag{1.19}$$

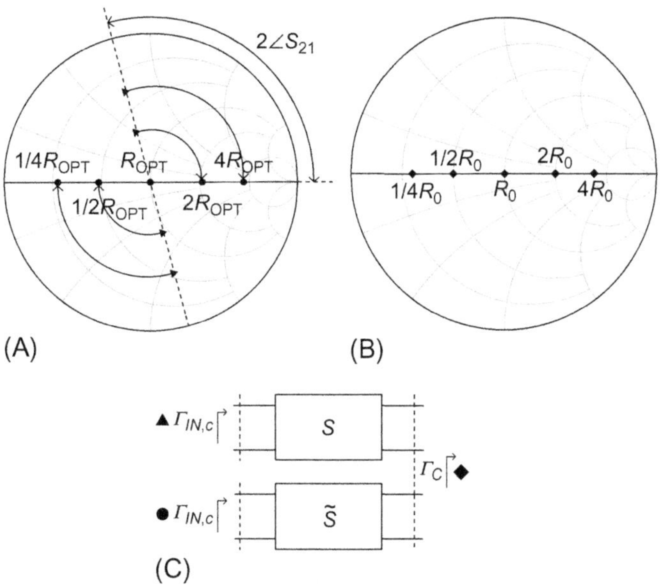


**Fig. 1.8** Load modulation without offset line (*triangle*) and with the offset line (*circle*): (A) $\Gamma_{IN,c}$ plane, (B) $\Gamma_c$ plane, and (C) *S*-parameters with/without the offset line. The same symbols are associated with the corresponding loads.

where $\Gamma_{C|R_0}$ is the reflection coefficient of $R_C$ with respect to $R_0$. Therefore, we can rewrite Eq.(1.18) as

$$\Gamma_{IN,c} = e^{j2\angle S_{21}}\Gamma_{C|R_0} \tag{1.20}$$

In other words, $\Gamma_{IN,c}$ corresponds to $\Gamma_{C|R_0}$ except for a phase factor of $e^{j2\angle S_{21}}$. As an example, the triangle symbols in Fig. 1.8 show the values of $\Gamma_{IN,c}$ corresponding to a possible load modulation of $\Gamma_c$ along the real axis. The trajectory of $\Gamma_{IN,c}$ still is a straight line but tilted with respect to the real axis following Eq. (1.20).

We insert an offset line with characteristic impedance $R_0$ and electric length $\theta_c = \angle S_{21}$ between the matching network and $R_C$. The modified network, represented by $\widetilde{S}$ (see Fig. 1.7), still matches $R_0$ to $R_{OPT}$, and the offset line introduces a delay $\theta_c$, compensating $e^{j2\angle S_{21}}$. Invoking the reciprocity of $\widetilde{S}$ and $S$, we have

$$\angle\widetilde{S}_{12} = \angle S_{12} - \theta_C = \angle S_{21} - \theta_C = \angle\widetilde{S}_{21} \tag{1.21}$$

Eq. (1.20) can therefore be rewritten as

$$\Gamma_{IN,c} = e^{j2\angle S_{21}}\Gamma_{C|R_0} = e^{j2(\angle S_{21} - \theta_C)}\Gamma_{C|R_0} \tag{1.22}$$

Hence, whatever load modulation acts on $R_C$, we set

$$\theta_c = \angle S_{21} + n\pi, \quad n \in \mathbb{N} \tag{1.23}$$

Then, the exactly same modulation is applied on $Z_{IN,c}$. In fact, in this case, we have

$$\Gamma_{IN,c} = \Gamma_{C|R_0} \Rightarrow \frac{Z_{IN,c}}{R_{OPT}} = \frac{R_C}{R_0} = \alpha \tag{1.24}$$

Fig. 1.8 (circle symbol) shows the behavior described by Eq. (1.24). When an offset line with electric length chosen according to Eq. (1.23) has been inserted, the modulation experienced by the reflection coefficient $\Gamma_C$ is now exactly reproduced on $\Gamma_{IN,c}$. It should be noticed that the impedance is calculated before the impedance inverter and $2R_0 \sim R_0$ load is converter to $2R_{OPT} \sim R_{OPT}$ at the current source.

If the offset line introduces a loss $A$, Eq. (1.24) can be rewritten as

$$\Gamma_{IN,c} = \Gamma_{C|R_0} e^{-2A} \Rightarrow \frac{Z_{IN,c}}{R_{OPT}} = \frac{\alpha + \tanh(A)}{1 + \alpha \tanh(A)} \tag{1.25}$$

In this case also, an offset line of length given in Eq. (1.23) ensures purely real load modulation but with reduced dynamic range when $R_C \neq R_0$.

#### *1.3.2.2 Offset Line at Peaking Amplifier*

Similar considerations can be applied to the offset line of the peaking amplifier, for which analogous expression of Eq. (1.18) can be derived from Fig. 1.7, where the parameters are changed to those of the peaking amplifier. Since the peaking amplifier should be open at a low power operation, the flow direction is reversed. But we can derive exactly the same equation as follows:

$$\Gamma_{OUT,p} = e^{j2\left(\angle S_{12}^{(P)} - \theta_P\right)} \frac{\Gamma_G - S_{11}^{(P)}}{1 - S_{11}^{(P)} \Gamma_G} \tag{1.26}$$

where $S^{(P)}$ represents the $S$-parameter of the cascaded circuit of the reactive elements and matching network of the peaking amplifier, the latter designed to ensure the output matching. $\Gamma_G$ is the peaking reflection coefficient similar to $\Gamma_C$. At a low-power drive, the peaking amplifier is turned off, and the intrinsic device behaves as an open circuit; thus,

$$\Gamma_{OUT,p}\big|_{\Gamma_G=1} = e^{j2\left(\angle S_{12}^{(P)} - \theta_P\right)} \tag{1.27}$$

From Eq. (1.27), it is clear that, without the offset line, the open circuit at the intrinsic device output plane is not correctly reproduced at the common load plane, while we obtain $\Gamma_{OUT,\ p} = 1$ with an offset line of length:

$$\theta_P = \angle S_{12}^{(P)} + n\pi, n \in \mathbb{N} \tag{1.28}$$

Since the matching network and reactive elements are composed of the reactive components, $S_{12}$ and $S_{21}$ have the same value. Therefore, the offset line for the peaking amplifier also assists for the load modulation like the offset line at the output of the carrier amplifier. With the offset line, the peaking amplifier properly operates; the output load of the peaking amplifier modulates from infinity to the $R_0$ while from infinity to $R_{OPT}$ at the current source of a device. Since the offset lines at the carrier and peaking amplifiers rotate the same angle to the real impedance region, under the condition that the device output capacitances and the output-matching circuits are the same, the lengths of the two offset lines should be identical. Also, it is worth to be noticed that, with help of the quarter-wave inverter at the carrier amplifier, the impedance trajectories for the carrier and peaking amplifiers are lined up but at the different impedance levels, i.e., $R_0 - R_0/2$ and $R_0 - \infty$, respectively.

## 1.4 OTHER LOAD MODULATION METHODS

So far, we have introduced the current-combining Doherty amplifier, which is the most popular design method in Doherty amplifiers. But there are several variances of the Doherty load modulation circuits. Doherty himself introduces a voltage-combining method also. In this section, we will introduce the different load modulation techniques of the Doherty amplifier.

### 1.4.1 Voltage Combined Doherty Amplifier

#### *1.4.1.1 Series Configured Doherty Amplifier in Voltage Combining Mode*

As Doherty described in his original paper, the Doherty load modulation circuit can be realized in a series voltage-combining mode of the two voltage sources. Fig. 1.9 shows the circuit schematic of the original Doherty amplifier with the series-connected load. The series-connected load properly combines the output powers delivered by the carrier and peaking amplifiers in a push-pull way since the load is ungrounded. It should be noticed that the impedance inverter is located at the peaking amplifier side and the load $R_L$ is $2R_{OPT}$ instead of $R_{OPT}/2$ in the current-mode case.

When the peaking amplifier is in off state with a high impedance, the load seen by the carrier amplifier is $R_L = 2R_{OPT}$ because the high impedance of the peaking amplifier is converted to a short by the inverter with $Z_T = R_{OPT}$. Conversely, when the carrier and peaking amplifiers deliver their full power, the two amplifiers equally share the $2R_{OPT}$

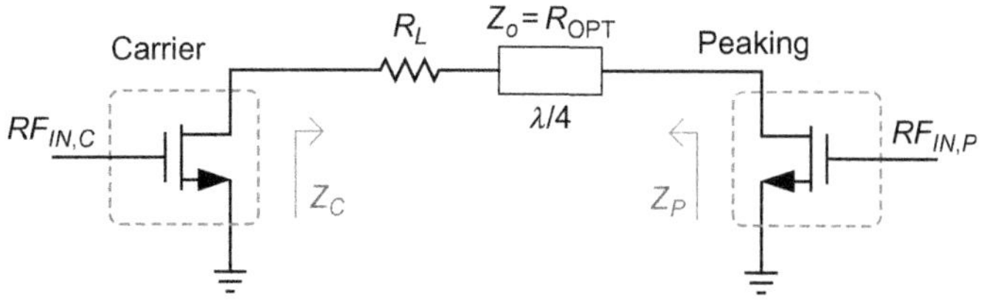


**Fig. 1.9** Series-configured Doherty amplifier in voltage-combining mode.

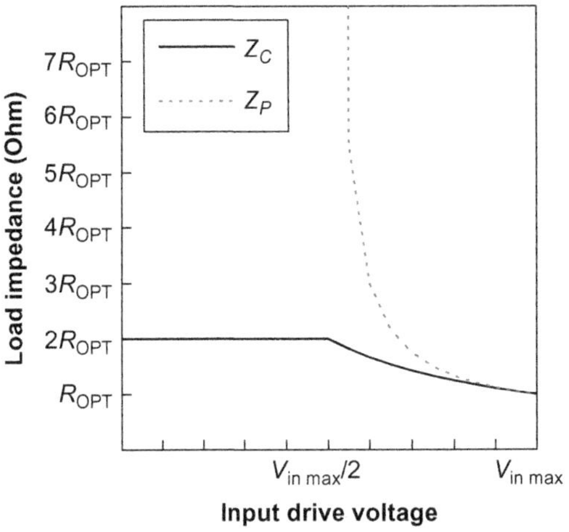


**Fig. 1.10** Load modulation behavior of the voltage-combined Doherty.

load. In between the two power levels, the loads seen by the carrier amplifier $Z_C$ and peaking amplifier $Z_P$ are modulated exactly the same way with current combining case. Since the Doherty modulation circuit provides the perfect matching, the calculation is simple.

$$Z_C = R_L - \frac{Z_O{}^2}{Z_P} = (R_L(I_1 - I_2') + V_2')/I_1 = 2R_{OPT}/(1+\alpha')$$
$$\text{where } V_2' = I_2' Z_P' = I_2'(R_L - Z_C) \tag{1.29}$$

$$Z_P' = \frac{R_L(I_2' - I_1) + V_1}{I_2'} = \frac{2R_{OPT}\,\alpha'}{(1+\alpha')}$$
$$\text{where } V_1 = I_1 Z_C = I_1(R_L - Z_P') \tag{1.30}$$

$$Z_P = \frac{Z_O{}^2}{R_L - Z_C} = \frac{Z_O{}^2}{Z_P'} = \frac{R_{OPT}(1+\alpha')}{2\alpha'}$$
$$\text{with } d = (I_2')/I_1 \tag{1.31}$$

where, $I_1$ and $V_1$ are the carrier current and voltage, $I_2$ and $V_2$ are the peaking current and voltage, $I_2'$ and $V_2'$ are the peaking current and voltage transferred by the inverter, respectively.

Thus, the load modulation behavior at the voltage sources is identical to the parallel configuration as shown in Fig. 1.10. Therefore, the operational behavior is also identical with the same voltage and current responses for the two sources. Since the powers generated from the two amplifiers are combined in series way, the $R_L$ is $2R_{OPT}$ instead of $R_{OPT}/2$ in the parallel case. It is evident that such a kind of configuration requires an output balun since the load is not connected to the ground, which is a big disadvantage of the series configuration.

#### *1.4.1.2 Transformer Based Power Amplifier*

The voltage-combining architecture is popular for a differential CMOS power amplifier with a transformer because the transformer can function as a voltage-combining balun. The differential structure is well suited for CMOS power amplifier to solve the low breakdown voltage and low power density of a CMOS device since the output load impedance is increased. The grounding problem of the CMOS power amplifier without via process is also solved by the differential structure. Therefore, a voltage-combined Doherty power amplifier based on the transformer is a natural choice for the CMOS process. The detailed load modulation behavior will be derived using the transformer-based CMOS Doherty power amplifier.

Fig. 1.11A shows a conceptual diagram of a conventional power amplifier architecture based on the transformer. $I_{PA}$ represents the current source of the amplifier. The transformer transforms the $50\,\Omega$ load impedance to a lower impedance at the primary for a power match of the amplifier using $1{:}m$ transformer. Based on the transformer theory,

$$V_2 = m \cdot V_1 \tag{1.32}$$

$$I_2 = \frac{1}{m} \cdot I_1 \tag{1.33}$$

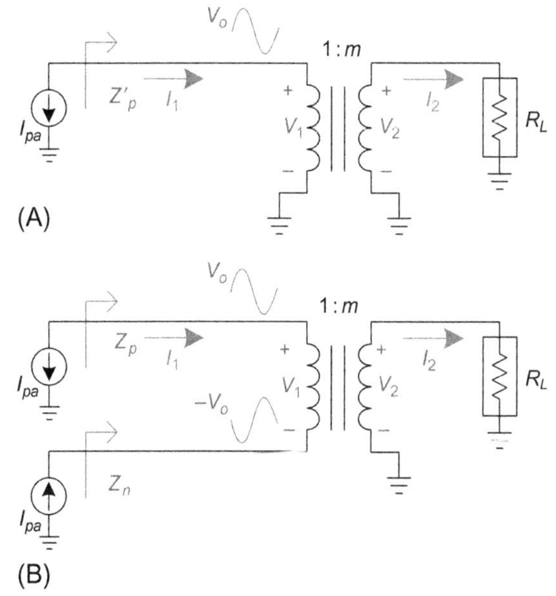


**Fig. 1.11** Conceptual operational diagram of transformer-based (A) single-ended and (B) differential power amplifiers.

In the differential power amplifier shown in Fig. 1.11B, the impedances are given by

$$R_L = \frac{V_2}{I_2} = \frac{2m^2 \cdot V_0}{I_1} \text{ (where } V_2 = 2m \cdot V_0) \tag{1.34}$$

$$Z_p = Z_n = \frac{V_0}{I_1} = \frac{R_L}{2m^2} \tag{1.35}$$

The optimum impedances $Z_p$ and $Z_n$ are inversely proportional to $2m^2$. However, in the single-ended power amplifier case shown in Fig. 1.11A, the impedances are

$$R_L = \frac{V_2}{I_2} = \frac{m^2 \cdot V_0}{I_1} \text{ (where } V_2 = m \cdot V_0) \tag{1.36}$$

$$Z'_p = \frac{V_0}{I_1} = \frac{R_L}{m^2} \tag{1.37}$$

The optimum impedance $Z'_p$ is two times larger, and the power cell size should be a half of the differential case. Therefore, the power level at the same matching impedance is 6 dB lower than that of the differential power amplifier. Therefore, the differential amplifier has a big advantage for power match of a high-power amplifier with a low impedance level.

#### *1.4.1.3 Transformer Based Voltage Combined Doherty Amplifier*

The transformer can modulate the load seen by the differential amplifiers, and this load modulation characteristic can be applied to the dynamic load modulation for a Doherty power amplifier. Fig. 1.12 shows the operational diagram of the voltage-combined Doherty power amplifier with a transformer. $I_C$ and $I_P$ are the current sources representing the carrier and peaking amplifiers, respectively, and their operations are identical to the current-combining case as described before. For the proper modulation with the voltage combining, the quarter-wave inverter is attached at the peaking

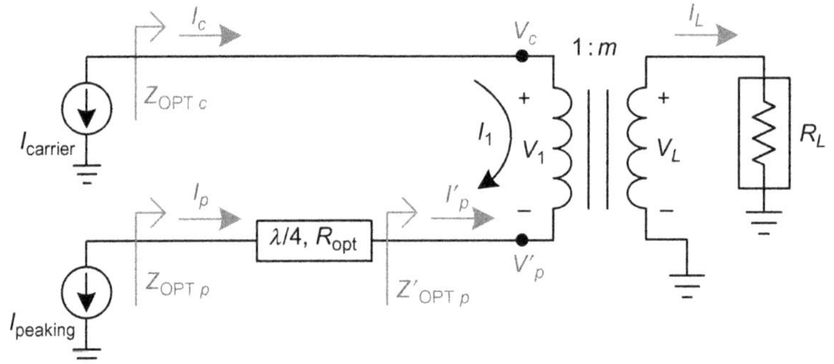


**Fig. 1.12** Operational diagram of voltage-combined Doherty power amplifier with a two-way transformer.

amplifier. When the peaking amplifier is turned off, the open impedance of the peaking is converted to the short at the transformer. Therefore, the carrier amplifier is connected to the $2R_{OPT}$ load instead of $R_{OPT}/2$ for the current-combining case (1:1 ratio case). When the two amplifiers generate their full powers, the two amplifiers share the load, $R_{OPT}$ each, and their powers are combined in the voltage-combining way. In between the two power levels, the transformer combines the two output powers properly as described below.

The modulation behavior can be calculated the same way as the current-combining case. It is assumed that each current source is linearly proportional to the input voltage after it is turned on, operating as a class B amplifier with harmonic short circuits. The peaking amplifier turns on at a half of the maximum input voltage. The phase difference between the carrier and peaking amplifiers is adjusted at the input. The two antiphased voltage signals are applied to the primary ports of the transformer, and the two signals are properly combined at the secondary output port. But, in this design, the carrier and peaking amplifiers do not operate in differential mode.

Fig. 1.13 shows the fundamental current and voltage amplitudes according to the input drive voltage. In the low-power region ($0 < V_{in} < V_{in,\ max}/2$), the peaking amplifier remains in the cutoff state, and the load impedance of the carrier amplifier is two times larger than that of the conventional amplifier. The operational behavior is the same as the single-ended operation explained earlier. The peaking amplifier's output impedance is infinity due to the open state. The quarter-wave inverter inserted at the output of the peaking amplifier transfers it to a short condition at the $V'_P$ node.

In the high-power region ($V_{in,\ max}/2 < V_{in} < V_{in,\ max}$), where the peaking amplifier is conducting, the load impedances of the carrier amplifier and peaking amplifier are modulated. Assuming that transconductance of the peaking amplifier is two times larger than that of the carrier amplifier, the current swings of the two amplifiers increase in

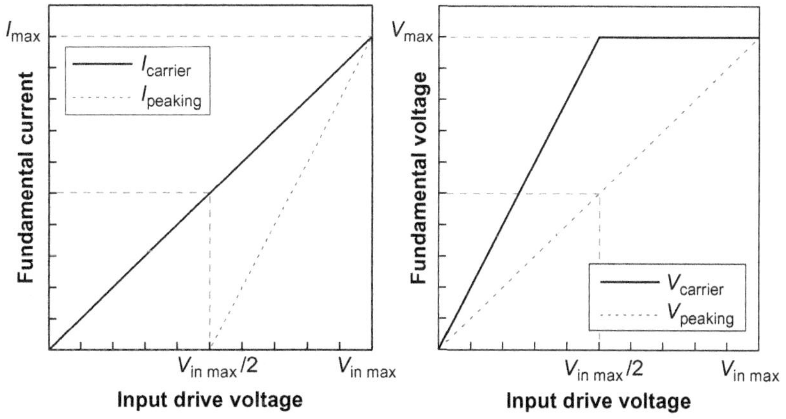


**Fig. 1.13** Magnitudes of the current and voltage versus the input drive voltage.

proportion to the input voltage level, and they are shown in Fig. 1.13A. The voltage swing of the peaking amplifier reaches to the $V_{max}$ only at the maximum input voltage while that of the carrier amplifier remains at $V_{max}$. This behavior can be explained using the modulated load impedances.

The load impedances seen by the carrier and peaking amplifiers can be calculated by using Eqs. (1.34), (1.35) with $V_0 = V_C - V'_P$ and given in Eqs. (1.38), (1.40), respectively (where $0 < V'_P < V_C$):

$$Z_{OPT,c} = \frac{V_C}{I_C} = \frac{R_L \cdot V_C}{m^2(V_C - V'_P)} (\text{where } I_C = I_1) \tag{1.38}$$

$$Z'_{OPT,p} = \frac{V'_P}{I'_P} = \frac{R_L \cdot V'_P}{m^2(V'_P - V_C)} (\text{where } I'_P = I_1) \tag{1.39}$$

$$Z_{OPT,p} = \frac{m^2 \cdot R^2_{OPT} \cdot (V'_P - V_C)}{R_L \cdot V'_P} \tag{1.40}$$

$Z_{OPT,p}$ is $Z'_{OPT,p}$ inverted by $R_{OPT}$ inverter. Therefore, the load impedances of the carrier and peaking amplifiers are dynamically modulated according to the input power level. Using Eqs. (1.38), (1.40) and the currents in Fig. 1.13A, the calculated voltage profiles are shown in Fig. 1.13B, where $R_L = 2R_{OPT}$ and $m = 1$ are assumed for a simple analysis (also a popular approach for a proper transformer design).

The magnitudes of the output currents $I_c$ and $I'_P$ at the primary of the transformer are the same with the $I_L$ at the secondary (where $I_c = -I_P = I_1 = I_L$) as shown in Fig. 1.14A. Since the currents are the same, the output voltages generated by the two amplifiers are voltage combined at the secondary of the transformer as shown in Fig. 1.14A. In the low-power region, the load impedance of $Z'_{OPT,p}$ at the combining node is maintained at zero, thanks to the quarter-wave inverter, as shown in Fig. 1.14C, and the carrier amplifier with the load impedance of $2R_{OPT}$ achieves a peak efficiency at 6 dB power back-off (PBO) point under the single-ended operation. In the high-power region, through Eqs. (1.38), (1.40), the transformer with the quarter-wave inverter properly modulates the load impedance seen by each carrier (from $2R_{OPT}$ to $R_{OPT}$) and peaking (from $\infty$ to $R_{OPT}$) amplifiers as shown in Fig. 1.14D. The load impedance of $Z'_{OPT,p}$ is inverted following Eq. (1.39).

As mentioned earlier, a differential operation is preferred for CMOS power amplifier. But the structure is not a differential mode and should be modified. A simple method is to operate the carrier and peaking amplifiers in differential mode, separately, and the four cells of the amplifiers are combined using a single transformer. Fig. 1.15 shows the circuit structure of the differential CMOS Doherty amplifier. The output power is four-way combined in this structure, which increases the output power by eight times for the same load impedance. Also, the imaginary part of the device impedance should be controlled

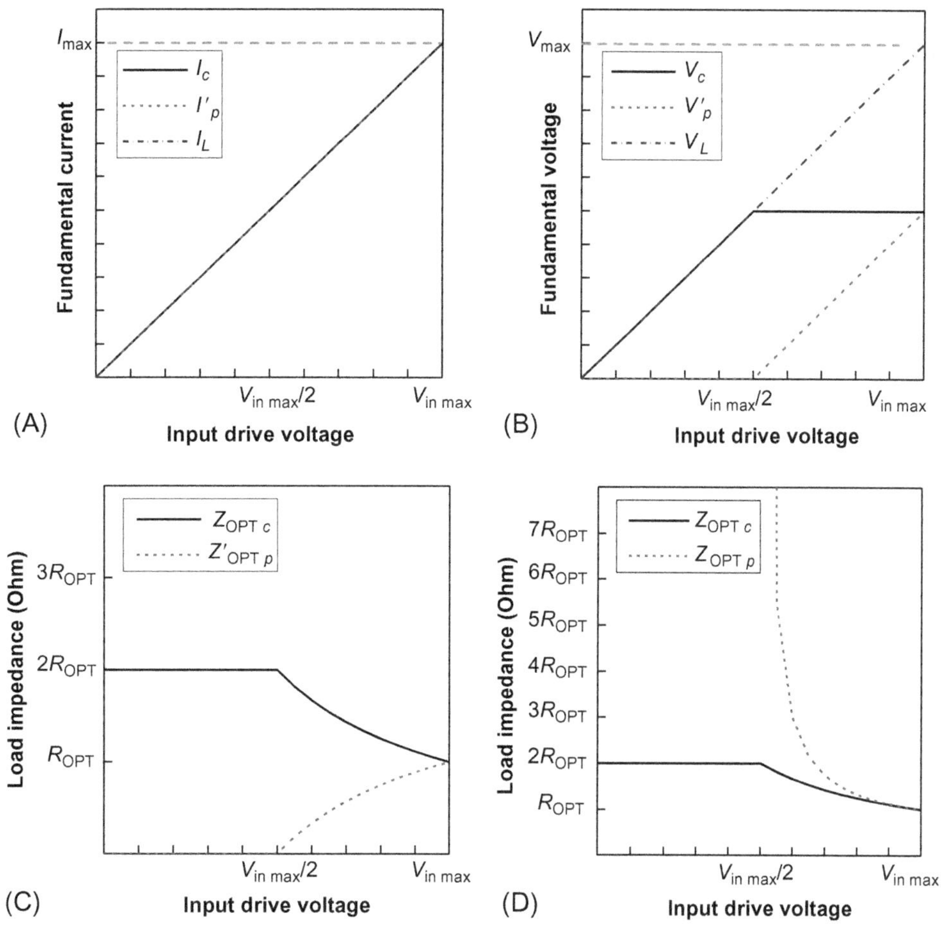


**Fig. 1.14** Operational behavior of the voltage-combining Doherty. (A) Current variation at the transformer, (B) voltage variation at the transformer, (C) load impedances at the primary ports of the transformer, and (D) load impedances at the current sources.

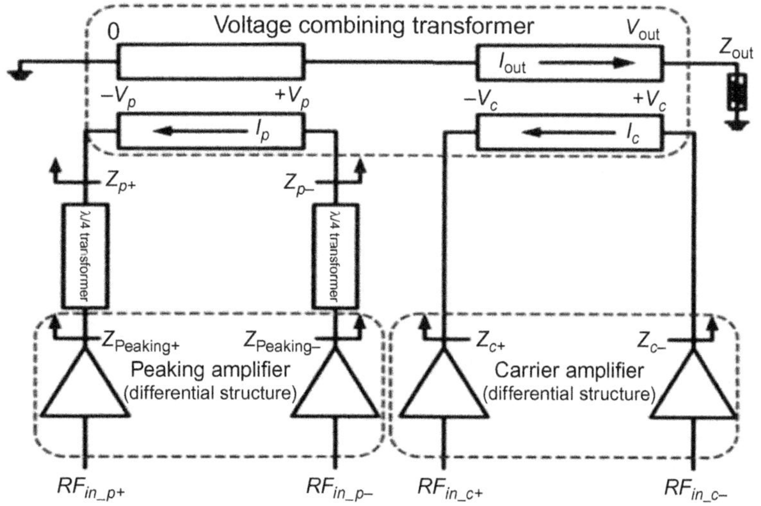


**Fig. 1.15** Block diagram of an ideal voltage-combining Doherty power amplifier with transformer combiner.

properly using an offset line or parallel resonation of the output capacitor at the drain terminal as we have described earlier.

## 1.4.2 Inverted Load Modulation

The inverted Doherty amplifier has the quarter-wave impedance inverter at the peaking amplifier instead of the carrier amplifier as shown in Fig. 1.16A. But it is a current-combining Doherty amplifier. This architecture is useful for the Doherty amplifier with the large capacitive output-matching impedances of the carrier and peaking amplifiers. In

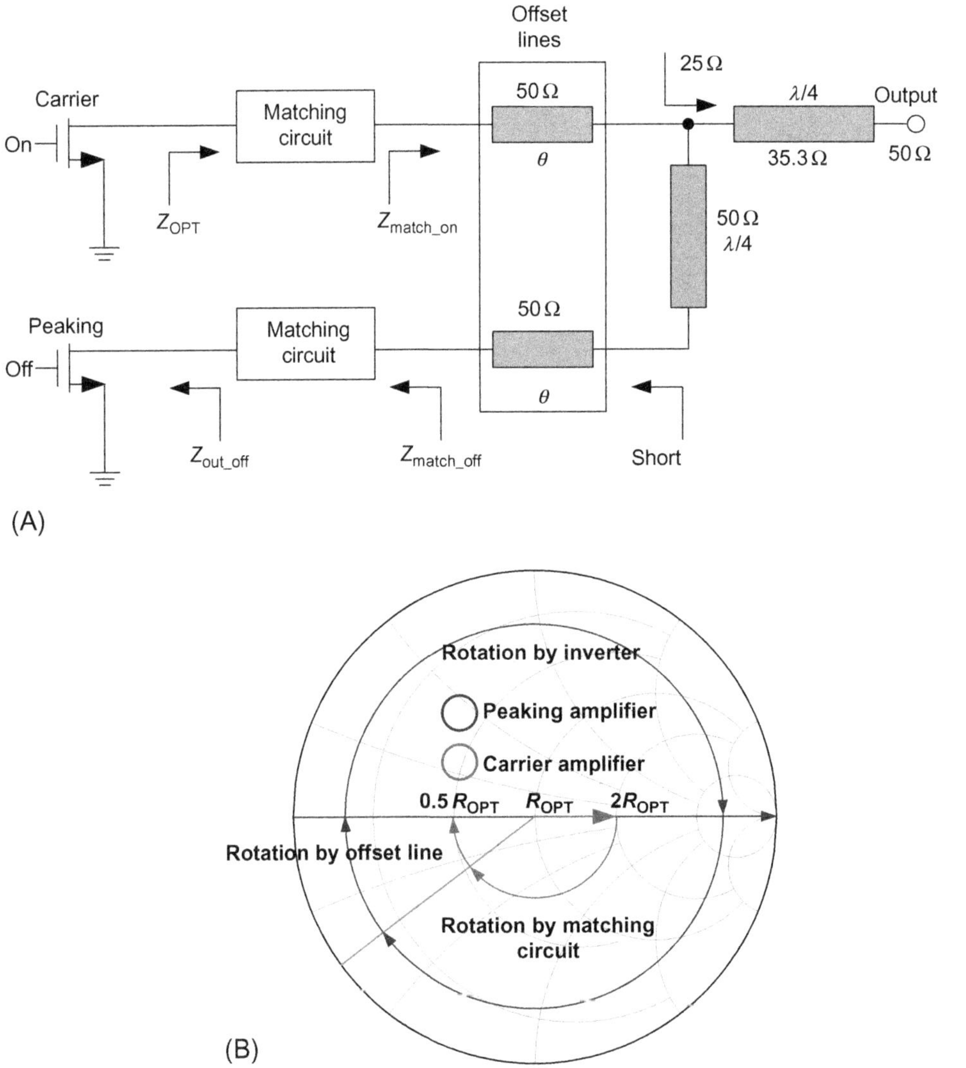


**Fig. 1.16** Load network: (A) schematic and (B) impedance rotations by the offset line and inverter.

this case, the length of the offset line can be shorter in the inverted Doherty architecture than the normal Doherty amplifier. But the behaviors of the two Doherty amplifiers are identical.

As discussed in Session 1.3, the load impedances of the carrier and peaking amplifiers are modulated from $R_L \sim R_L/2$ and $R_L \sim \infty$, respectively. The $R_L \sim R_L/2$ at the carrier amplifier is converted to $R_L \sim 2R_L$ by the $R_L$ line inverter. The impedances should be transferred to $R_{OPT} \sim 2R_{OPT}$ and $R_{OPT} \sim \infty$ at the current sources of the carrier and peaking amplifiers, respectively. But the impedances are rotated by the output capacitances and matching circuits as we discussed earlier. The rotation angles for the two amplifiers can be identical under the assumption that they are identical amplifiers without their output capacitance variation. Therefore, by the same offset line, the impedance lines can be rotated to the real axis with the proper values.

The load impedance lines for the carrier and peaking amplifiers located at the capactive region are depicted in Fig. 1.16B. In this case, as shown in Fig. 1.16B, the lines are rotated to the resistive axis by the short offset line. For the carrier amplifier, the load after the offset line is transformed to $R_{OPT} \sim 0.5R_{OPT}$, without the help of the impedance inverter, for the load impedance change of $R_L \sim 2R_L$. For the peaking amplifier, the line at $R_{OPT}\sim$ short should be rotated further to $R_{OPT} \sim \infty$ region by the inverter located the peaking amplifier. Therefore, this structure requires the offset line at the peaking amplifier but the operational behavior is identical to the conventional Doherty amplifier.

In comparison, the conventional Doherty amplifier requires the offset line that is a quarter-wave length longer than this inverted Doherty amplifier since the line should be rotated to $R_{OPT} \sim 2R_{OPT}$. Then the inverter rotates the line by 180°, reaching to $R_{OPT} \sim 0.5R_{OPT}$. However, the peaking amplifier with the longer offset line does not need the inverter. Therefore, the total length of the line at the carrier amplifier is two quarter-wave lengths shorter for the inverted structure. However, when the matching impedance lines are located at the inductive region, the conventional Doherty amplifier requires shorter offset lines. Therefore, we can select a better topology between the conventional and inverted load networks after considering $Z_{match\_on}$ for the fully matched carrier amplifier. If the $Z_{match\_on}$ of the carrier amplifier is located from 0° to −180°, we have to choose an inverted load network that requires shorter offset lines. Otherwise, the conventional Doherty network requires shorter offset lines. Another solution is, since the high-pass-type filter produces a negative angle rotation, the high-pass-type offset line can be employed instead of the inverter Doherty topology, if the filter is available like an on-chip circuit.

### 1.4.3 Direct Matching at the First Peak Efficiency Point

For amplification of a modulated signal with a large PAPR, the first peak efficiency region is more important than the high-power region because major portion of the amplification is carried out at the first peak region. Therefore, it is advantageous to design the Doherty amplifier with accurate match at the first peak efficiency region. However, the conventional Doherty amplifier is directly matched at the peak-power point by the matching circuit without being affected by the load modulation. And the impedance at the first peak efficiency region is matched to $2R_{OPT}$ through the load modulation. But it is difficult to accurately match at the first peak region in this design because the load modulation does not behave ideally due to the nonlinear device characteristics.

Therefore, it is advantageous to match directly at the first peak because the modulation circuit sensitivity is moved to the peak power region, relaxing the circuit tolerance at the first peak matching. Moreover, at the peak-power region, the carrier and peaking amplifiers are operated at the saturated mode, where the circuit tolerance is inherently large. The direct matching at the first peak can be done in two ways.

#### *1.4.3.1 Using $R_{OPT}/2$ Inverter*

In a conventional Doherty amplifier, the output-matching circuit of the carrier amplifier is designed to match the optimum output impedance of $R_{OPT}$ at the peak output power region to $R_0$, and the characteristic impedances of the offset line and quarter-wave inverter are set to the same impedance, $R_0$, as depicted in Fig. 1.17A. The offset line and inverter modulate the load for matching to $2R_{OPT}$ at the back-off power region, while the match at the peak-power operation is not affected by the modulation circuit. In this design, the accurate output matching at the first peak efficiency region is very difficult due to the nonlinear load modulation behavior of the amplifier with a low matching tolerance. The load modulation does not behave ideally either due to the nonlinear device parameters variation. However, to get a high efficiency for amplification of a modulated signal, the accurate match at the back-off power region is more important than that at the peak-power region.

To solve the problem, the output-matching circuit of the carrier amplifier is matched at the output power level with the first peak efficiency, from $2R_{OPT}$ optimum load to $R_0/2$ load. The impedance inverter and offset line of the carrier power amplifier are also designed with the characteristic impedance of $R_0/2$, and the match is not affected by the load modulation circuit (Fig. 1.17B). The load is modulated for match at the high-power region. The new offset line compensates the reactive element effect for matching from $R_{OPT}$ to $R_0/4$ load as shown in Fig. 1.18, and the $R_0/4$ load is inverted to $R_0$. And the load modulation behavior of the carrier amplifier is the same as the conventional design.

We can design the peaking amplifier matched at the first peak region, similarly to the carrier amplifier. But the peak-power region is more important for the peaking amplifier since no power is generated at the first peak region and the peaking amplifier is designed the same way as the conventional case.

Fig. 1.19 shows the output impedance trajectory of the carrier amplifier through the load modulation, which is simulated with a Cree gallium nitride (GaN) HEMT

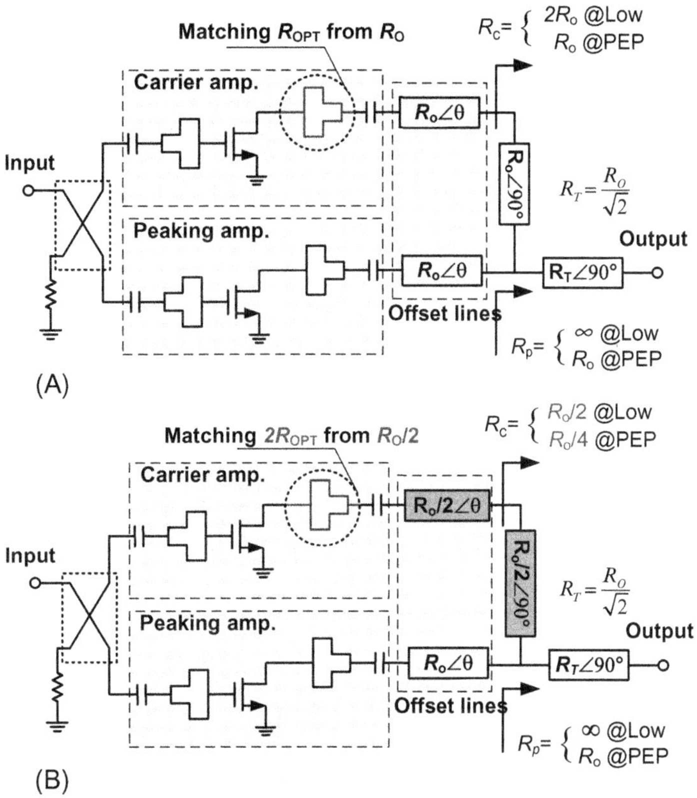


**Fig. 1.17** Schematics of the Doherty amplifiers: (A) the conventional Doherty amplifier and (B) the first peak-matched Doherty power amplifier.

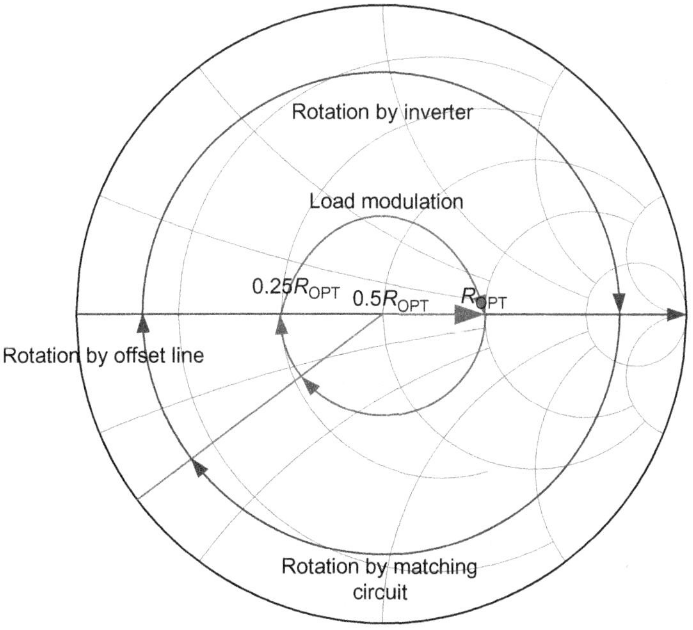


**Fig. 1.18** Load modulation of carrier amplifier with $R_O/2$ inverter and offset line.

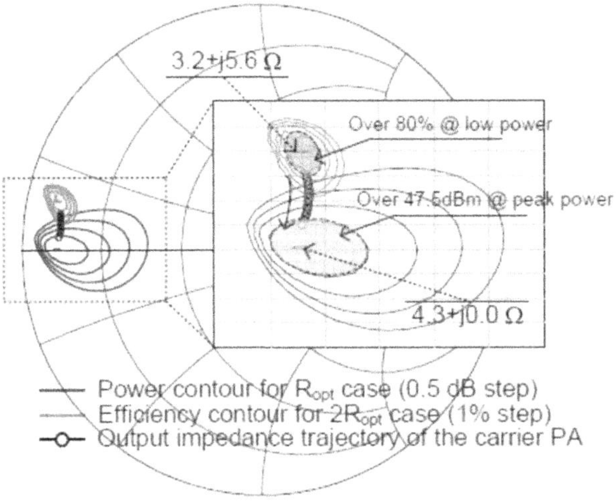


**Fig. 1.19** Results of load-pull simulation and output impedance trajectory of the carrier amplifier through the load modulation.

CGH40045 model. In the Doherty amplifier design, the efficiency at the first peak region is very important because overall efficiency is determined by the efficiency, while the power is important at the peak-power region because the power limits the dynamic range of the load modulation while the efficiency does not affect the overall efficiency significantly. Therefore, the carrier power amplifier is efficiently matched to $R_0/2$ at the back-off power with $2R_{\mathrm{OPT}}$, and the offset line is adjusted to deliver the maximum output power at the peak power. As shown in Fig. 1.19, this design concept of the first peak match has advantage in realization because the power contour at the peak-power region has a large tolerance of the matching impedance and the carrier power amplifier can be matched accurate to the high efficiency region at the low power region with low tolerance.

#### *1.4.3.2 Using 2R$_{OPT}$ Inverter*

The other method is the use of the inverter with

$$Z_T = R_L = 2R_0 \tag{1.41}$$

Here, the carrier and peaking amplifiers are matched to $2R_0$ load from $2R_{\mathrm{OPT}}$ device impedance at the first peaking efficiency region. In this Doherty amplifier design shown in Fig. 1.20, the output impedance of the carrier amplifier $Z_m$ is given by

$$Z_m = \frac{V_c}{I_c} = {Z_T}^2 \cdot \frac{I_T}{V_L} = {Z_T}^2 \cdot \frac{I_L - I_a}{V_L} = \left(\frac{Z_T}{R_L} - \frac{I_a}{I_C}\right) \cdot Z_T \tag{1.42}$$

$$I_T = \frac{I_m \cdot Z_m}{Z_T} = \frac{I_m \cdot Z_T}{Z_L} \quad \text{with} \quad Z_L = \frac{V_L}{I_T} \tag{1.43}$$

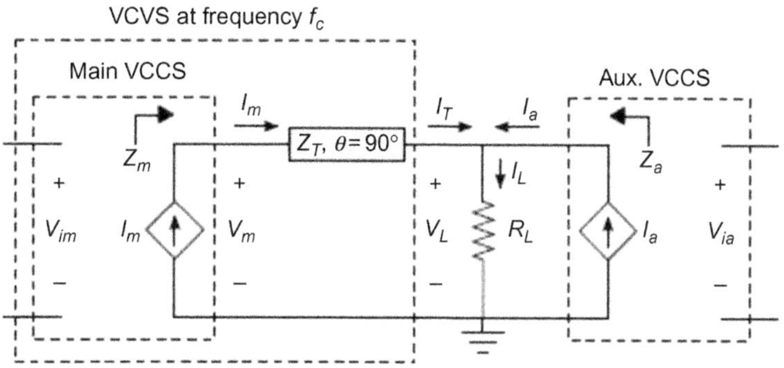


**Fig. 1.20** The first peak-matched Doherty amplifier using $2R_0$ inverter.

$$Z_a = V_L/I_a = I_m \cdot Z_T/I_a \tag{1.44}$$

At the peak-power operation, the impedance of the carrier amplifier is $R_{OPT}$, and this impedance is transferred to $4R_0$ by the inverter with $Z_T = 2R_0$. Therefore, $I_T$ given by Eq. (1.43) should be identical to $I_a$ and is given by

$$I_T = I_m/2 = I_a \tag{1.45}$$

With Eqs. (1.44, (1.45), the $I_m$ and $I_a$ profiles versus the input voltage $V_{in}$ are shown in Fig. 1.21, which are mathematically given as

$$I_m = \frac{V_{in}}{V_{in,max}}(I_{max}/2) = v_{in}(I_{max}/2) \tag{1.46}$$

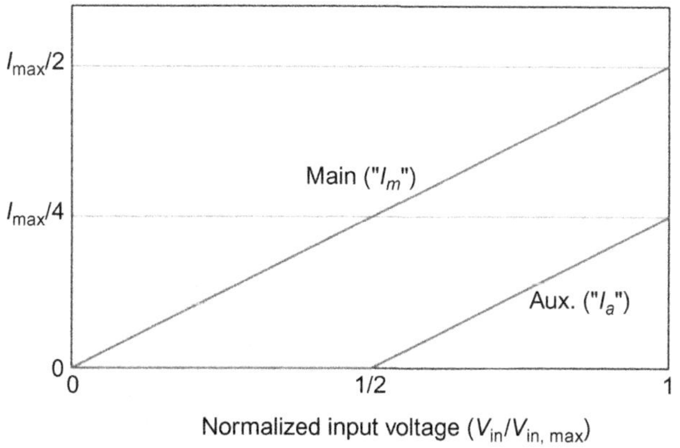


**Fig. 1.21** Current profiles of the carrier amplifier and peaking amplifier versus the normalized input voltage.

$$I_a = \begin{cases} 0, & 0 \leq v_{in} < 0.5 \\ \left(v_{in} - \frac{1}{2}\right)\left(\frac{I_{max}}{2}\right), & 0.5 \leq v_{in} \leq 1 \end{cases} \tag{1.47}$$

The $I_a$ versus $V_{in}$ function in Eq. (1.47) can be easily realized in practice using a peaking device with the same size as the main device, except biased in class C. But only a half of the current capability of the peaking amplifier is utilized.

$Z_m$, the impedance seen by the carrier device, is identical to that of the conventional Doherty with perfect tracking of $2R_{OPT}$ for up to 6 dB of back-off power. At the peak power, the peaking device now sees $4R_{OPT}$ instead of $R_{OPT}$ because $V_L$ is doubled while $I_a$ is halved. Fig. 1.22 shows the load modulation in the Doherty amplifier calculated using Eqs. (1.42) and (1.44).

The voltages across the carrier and peaking devices, $V_m$ and $V_a$, can be calculated using (1.42) and (1.46), (1.44) and (1.47), respectively. The voltage profile is shown in Fig. 1.23. As shown, the $V_m$ is identical to that of the conventional Doherty amplifier in Fig. 1.3 in Section 1.3. On the other hand, the voltage $V_L$ across the peaking device and the load now swings twice as much as the $V_L$ of the conventional Doherty amplifier in the figure. As such, the DC drain bias of the peaking amplifier needs to be twice that of the carrier amplifier. From a practical perspective, the need for asymmetrical bias voltages implies that only a half of power capability of the carrier device is utilized for the power generation.

The efficiency curve and the peak output power are identical to that of the conventional Doherty power amplifier as shown in Fig. 1.24. This design method can achieve the goal of the accurate first peak match. But the carrier amplifier utilizes a half of the

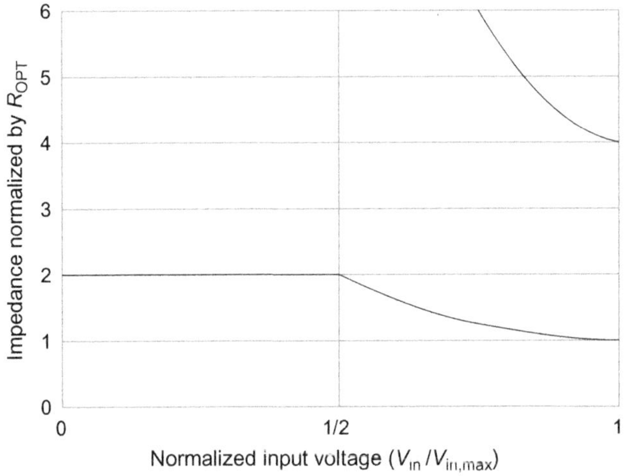


**Fig. 1.22** Calculated impedances $Z_m$ and $Z_a$ versus the normalized input voltage.

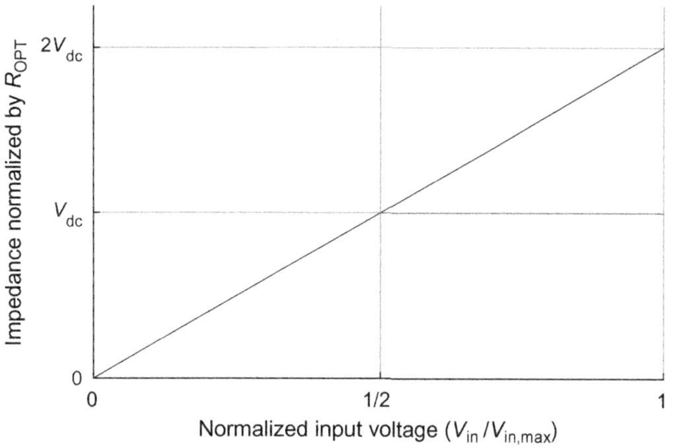


**Fig. 1.23** Calculated voltages across the carrier amplifier and the voltage across the load or peaking amplifier versus the normalized input voltage.

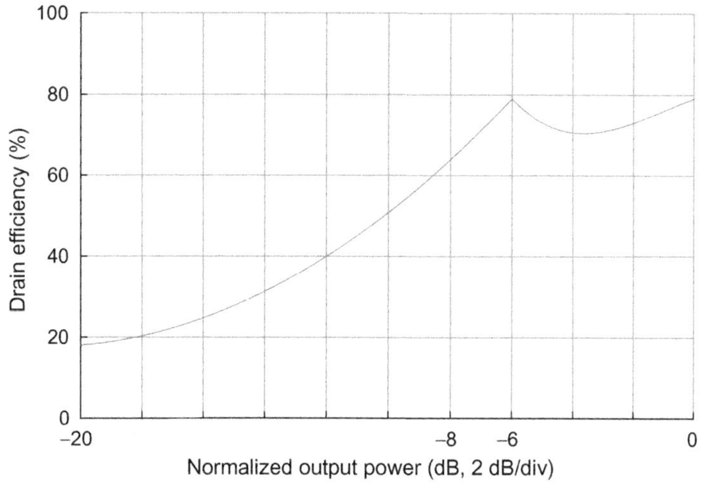


**Fig. 1.24** The calculated drain efficiency versus normalized output power in the proposed Doherty amplifier.

available voltage swing and the peaking amplifier a half of the current maximum current-handling capability. Those problems are serious limitation for this design technique. As we mentioned earlier, the peaking amplifier in the conventional Doherty amplifier utilizes a half of the current available, also. This problem can be solved by uneven drive that will be discussed in the next chapter.

## FURTHER READING

[1] R. Pengelly, et al., Doherty's legacy: a history of the Doherty power amplifier from 1936 to the present day, IEEE Microw. Mag. 17 (2) (2016) 41–58.
[2] Y. Yang, et al., Optimum design for linearity and efficiency of microwave Doherty amplifier using a new load matching technique, Microw. J. 44 (12) (2001) 20–36.
[3] B. Kim, et al., The Doherty power amplifier, IEEE Microw. Mag. 7 (5) (2006) 42–50.
[4] R. Quaglia, et al., Offset lines in Doherty power amplifiers: analytical demonstration and design, IEEE Microwave Wireless Compon. Lett. 23 (2) (2013) 93–95.
[5] Y. Cho, et al., Voltage-combined CMOS Doherty power amplifier based on transformer, IEEE Trans. Microwave Theory Tech. 64 (11) (2016) 3612–3622.
[6] S. Kwon, et al., Inverted-load network for high-power Doherty amplifier, IEEE Microw. Mag. 10 (1) (2009) 93–98.
[7] Y. Park, et al., Optimized Doherty power amplifier with a new offset line, IEEE MTT-S Int. Microw. Sympo. Dig., Phoenix, AZ, May, 17–22, 2015.
[8] D. Yu-Ting Wu, S. Boumaiza, A modified Doherty configuration for broadband amplification using symmetrical devices, IEEE Trans. Microwave Theory Tech. 60 (10) (2012) 3201–3213.

CHAPTER TWO

# Realization of Proper Load Modulation Using a Real Transistor

In this chapter, we will introduce various techniques to compensate the nonideal operation of the Doherty amplifier related to the transistor characteristics. For the proper load modulation of the Doherty amplifier, transconductance of the peaking amplifier should be two times larger than that of the carrier amplifier. Therefore, the transistor of the peaking amplifier should be twice bigger than that of the carrier, but only a half of the full current capability is utilized, wasting the precious transistor resource. This problem can be solved by the uneven input power drive, 6 dB more input drive to the peaking amplifier. For the same purpose, the gate bias of the peaking amplifier can be adapted, increased the gate bias, as a square root function of input power, from the pinch-off point to the same bias with the carrier amplifier at the peak power operation. Ideally, these techniques provide the proper Doherty load modulation. However, the load modulation can be deviated from the ideal case due to the knee voltage effect of a transistor and the large input capacitance variation of the class-C-biased peaking amplifier. In this chapter, the remedies for the nonideal operations are described.

## 2.1 CORRECTION FOR LOWER CURRENT OF PEAKING AMPLIFIER

Due to the class C bias and the two times larger transconductance required for the peaking amplifier, the fundamental component of the current generated by the peaking amplifier, with the same size transistor as the carrier amplifier, is always lower than the required level. Therefore, the load impedances of the carrier and peaking amplifiers cannot be modulated fully to the optimum impedance levels, and the two amplifiers see larger loads than the ideal cases. The most significant deviation is occurred at the high-power region. The carrier amplifier operates in a saturated mode at the modulated region due the higher load impedance, producing a large distortion. The resulting Doherty amplifier produces a lower output power than the anticipated full power, but the efficiency can be high due to the saturated operation. This problem can be solved by using a peaking amplifier with twice bigger size than that of the carrier amplifier. In this design, only a half of the full power capability of the peaking device is utilized, increasing the production cost. But the low fundamental current generation in the class C operation cannot be fixed completely. To solve the problem, the uneven input power drive or gate-bias adaptation technique is suggested.

In the uneven power drive, the peaking amplifier gets more input power than the carrier amplifier to compensate its lower bias. In this drive, the currents of both amplifiers

*Doherty Power Amplifiers*
https://doi.org/10.1016/B978-0-12-809867-7.00002-8

can be the same at the full power level. As a result, the impedances of both amplifiers are fully modulated, and the optimum power matching is achieved at the high power level, delivering the maximum power to the common load. Since the input power to the carrier amplifier is reduced, however, the gain of the uneven drive Doherty amplifier is lower.

The uneven drive can be realized by using an uneven power divider or direct power dividing using the large input capacitance variation of the class-C-biased peaking amplifier. The coupler provides an isolation between the carrier and peaking amplifiers, and it is more stable. Therefore, the coupler is popularly employed in base-station amplifier. The direct dividing is a simpler method without using the bulky coupler and is a favored choice for handset Doherty amplifiers. This direct dividing technique is described in handset Doherty power amplifier chapter (Section 5.3.1).

Gate-bias adaptation of the Doherty amplifier is another approach to solve the problem. As the magnitude of the instantaneous input signal increases, the gate-bias voltage of the peaking amplifier increases according to the envelope of the signal. Thus, the peaking amplifier generates the proper current for the required Doherty load modulation. However, it needs an external voltage control circuitry, which adds the circuit complexity. Since the bias voltage is modulated at the signal envelope frequency speed, the envelope frequency harmonics generated by the amplifiers cannot be suppressed using a large capacitor at the gate, creating the memory effect. This problem can be solved by using a voltage control circuit with a low output impedance. The circuit operates at a low power level for the gate-bias control, and the efficiency of the amplifier is not affected.

## 2.1.1 Uneven Drive Through Coupler

### *2.1.1.1 Current Ratio of Peaking Amplifier Versus Carrier Amplifier*

Doherty amplifier consists of a class-AB-biased carrier amplifier and a class-C-biased peaking amplifier. Since both amplifiers have identical size transistors and input drives, the current of the peaking amplifier is always lower than the required current level, and the current at the maximum input drive reaches to far below the required maximum current level. Fig. 2.1 illustrates the fundamental and harmonic current levels of a general amplifier as a function of conduction angle. In Fig. 2.1, the current level is normalized to the maximum channel current, and the operation regions for classes AB and C are also indicated. As shown, the fundamental component of the current $I_1$ is limited to

$$I_1 = \begin{cases} 0.5 \sim 0.536, & \alpha = \pi \sim 2\pi \\ 0 \sim 0.5, & \alpha = 0 \sim \pi \end{cases} \tag{2.1}$$

where $\alpha$ is the conduction angle. The fundamental components of the carrier and peaking currents at the maximum drive $I_{1,\ C}$ and $I_{1,\ P}$ are given by

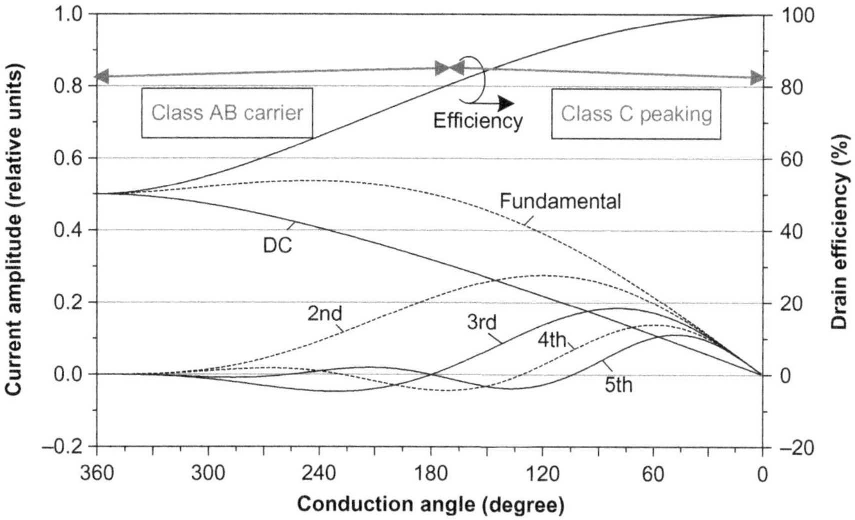


**Fig. 2.1** Fundamental and harmonic components of the conduction current versus the conduction angle for a general amplifier.

$$I_{1,C} = \frac{I_{\max}}{2\pi} \cdot \frac{\alpha_c - \sin\alpha_C}{1 - \cos\left(\frac{\alpha_c}{2}\right)} \tag{2.2}$$

$$I_{1,P} = \frac{I_{\max}}{2\pi} \cdot \frac{\alpha_p - \sin\alpha_p}{1 - \cos\left(\frac{\alpha_p}{2}\right)} \tag{2.3}$$

where $\alpha_c$ and $\alpha_p$ are the conduction angles of the carrier amplifier and peaking amplifier, respectively. Due to the class C bias, the peaking amplifier is not fully driven even at the maximum input drive for the carrier amplifier, and the current level of the peaking amplifier is reduced further by a factor of $(1-N)$, where $N$ denotes the portion of the voltage where the peaking amplifier starts to conduct. Here, $\alpha_p$ also changes according to $N$.

We can define the current ratio of the two amplifiers at the maximum drive for the carrier amplifier as

$$\sigma = \frac{I_{1,C}}{I_{1,P} \cdot (1-N)} \tag{2.4}$$

The output power ratio from the two amplifiers in decibel, $E_I$, is given by

$$E_I = 10\log\left(\sigma^2\right) = 10\log\left(\frac{I_{1,C}}{I_{1,P}}\right)^2 + 10\log\left\{\frac{1}{(1-N)^2}\right\} \tag{2.5}$$

Note that $E_I$ is the criterion to determine the underdrive ratio of the peaking amplifier. To generate the same current, the peaking amplifier should be driven harder than the carrier amplifier by that ratio. For example, if $I_{1,C} = 0.52$ ($\alpha_c = 200°$) and $I_{1,P} = 0.44$ ($\alpha_p = 140°$), the load could be modulated fully by increasing the drive power for the peaking amplifier (in this case, $N$ is 0.5) by 7.6 dB more than the carrier amplifier, as determined from Eq. (2.5) with $\sigma = 2.4$.

### 2.1.1.2 Efficiency of the Asymmetric Amplifier With Uneven Power Drive

Fig. 2.2A shows an operational diagram to analyze the efficiency of the amplifier with an uneven drive, $\sigma^2$ times larger drive to the peaking amplifier. It is assumed that each

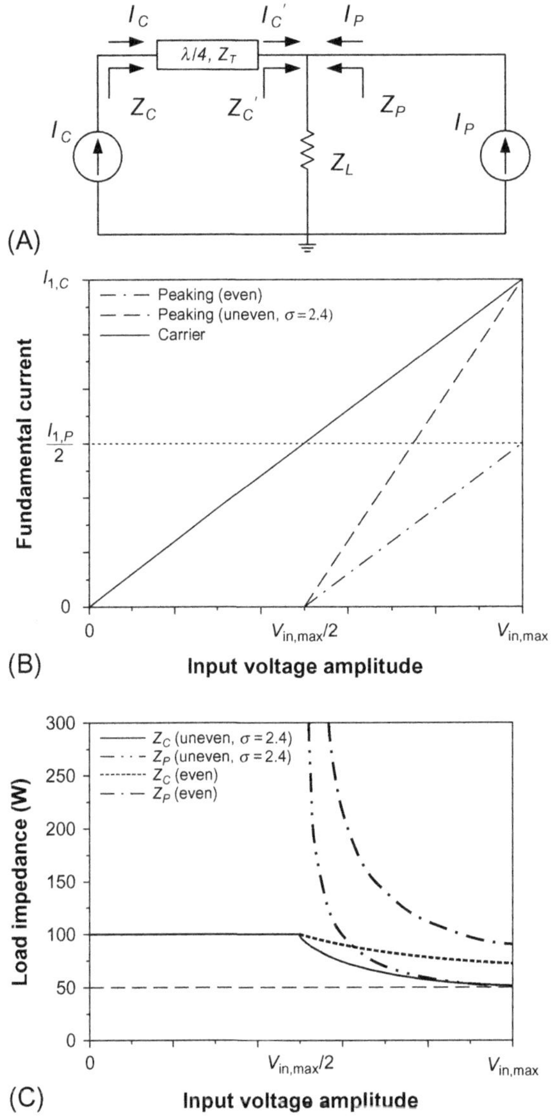


**Fig. 2.2** (A) Operational diagram with uneven power drive. (B) Fundamental currents versus input voltage. (C) Load impedances versus input drive for even and uneven ($\sigma = 2.4$) drives (carrier ($\alpha_c = 200^\circ$), peaking ($\alpha_p = 140^\circ$), $N = 0.5$, $Z_o = 50\ \Omega$, and $Z_T = 50\ \Omega$).

current source is linearly proportional to the input voltage signal and they contain ideal harmonic short circuits so that the efficiency analysis can be carried out using the fundamental and dc components only. Fig. 2.2B shows the fundamental currents from the two amplifiers for the even and uneven ($\sigma = 2.4$) drives. As shown, the carrier amplifier is biased at a class AB and the peaking amplifier at C, turning on at a half of the maximum voltage ($V_{\text{in,max}}$). As shown in Fig. 2.2C, we can achieve the proper load modulation suggested by Doherty with the uneven power drive since the two amplifiers generate the proper current with the same maximum value.

From Fig. 2.2B, the currents from the carrier and peaking amplifiers are given by

$$I_C = \frac{I_{1,C}}{V_{\text{in,max}}} \cdot v_{\text{in}}, \quad 0 < v_{\text{in}} < V_{\text{in,max}} \tag{2.6}$$

$$I_{P,\text{even}} = \begin{cases} I_{1,P} = 0, & 0 < v_{\text{in}} < N \cdot V_{\text{in,max}} \\ \left(\dfrac{I_{1,P}}{V_{\text{in,max}}}\right) \cdot v_{\text{in}} - N \cdot I_{1,P}, & N \cdot V_{\text{in,max}} < v_{\text{in}} < V_{\text{in,max}} \end{cases} \tag{2.7}$$

$$I_{P,\text{uneven}} = \begin{cases} \sigma I_{1,P} = 0, & 0 < v_{\text{in}} < N \cdot V_{\text{in,max}} \\ \left(\dfrac{\sigma I_{1,P}}{V_{\text{in,max}}}\right) \cdot v_{\text{in}} - N \cdot \sigma I_{1,P}, & N \cdot V_{\text{in,max}} < v_{\text{in}} < V_{\text{in,max}} \end{cases} \tag{2.8}$$

where $I_{P,\text{even}}$ is the current level of the peaking amplifier with the even power drive and $I_{P,\text{uneven}}$ is the current level of the peaking amplifier with the uneven power drive.

The load impedances of the two amplifiers are given by

$$Z_C = \begin{cases} \dfrac{Z_T^2}{Z_L}, & 0 < v_{\text{in}} < N \cdot V_{\text{in,max}} \\ \dfrac{Z_T^2}{\left[Z_L \cdot \left(1 + \dfrac{I_P}{I_C'}\right)\right]}, & N \cdot V_{\text{in,max}} < v_{\text{in}} < V_{\text{in,max}} \end{cases} \tag{2.9}$$

$$Z_P = \begin{cases} \infty, & 0 < v_{\text{in}} < N \cdot V_{\text{in,max}} \\ Z_L\left(1 + \dfrac{I_C'}{I_P}\right), & N \cdot V_{\text{in,max}} < v_{\text{in}} < V_{\text{in,max}} \end{cases}. \tag{2.10}$$

At the low-power region ($0 < v_{\text{in}} < N \cdot V_{\text{in,max}}$), where only the carrier amplifier is active, the RF and dc powers increase according to the input drive voltage. If we use $Z_T = Z_o$ and $Z_L = Z_o/2$, the RF power and dc power are given by

$$P_{RF} = \frac{1}{2} I_C^2 Z_C = \frac{1}{2}\left(\frac{v_{\text{in}} \cdot I_{1,C}}{V_{\text{in,max}}}\right)^2 \cdot 2Z_0 = I_{1,C}\left(\frac{v_{\text{in}}}{V_{\text{in,max}}}\right)^2 \cdot V_{dc} \tag{2.11}$$

$$P_{dc} = I_{dc,C} \cdot V_{dc} = I_{dc\,\max,C}\left(\frac{v_{\text{in}}}{V_{\text{in,max}}}\right) \cdot V_{dc} \tag{2.12}$$

where $V_{dc}$ is the drain bias voltage of the two amplifiers. From Eqs. (2.11), (2.12), the efficiency becomes

$$\eta = \frac{P_{RF}}{P_{dc}} \times 100 = \frac{I_{1,C}}{I_{dc,C}}\left(\frac{v_{\text{in}}}{V_{\text{in,max}}}\right) \times 100 \tag{2.13}$$

At the higher-power region ($N \cdot V_{\text{in,max}} < v_{\text{in}} < V_{\text{in,max}}$), both amplifiers are active. Thus, the total RF and dc powers are obtained by adding those of the two amplifiers and are given by Eqs. (2.14), (2.15).

$$\begin{aligned} P_{RF} &= \frac{1}{2}\left(I_C^2 Z_C + I_P^2 Z_P\right) \\ &= V_{dc} \cdot \left[\frac{I_{1,C}^2 \cdot V^3}{(I_{1,C} + \sigma I_{1,P}) \cdot V - N \cdot \sigma I_{1,P}} + \frac{\sigma I_{1,P}(V-N)\{(I_{1,C} + \sigma I_{1,P})V - N \cdot \sigma I_{1,P}\}}{4 I_{1,C}}\right] \end{aligned} \tag{2.14}$$

$$P_{dc} = \left(I_{dc,C} + I_{dc,P}\right) \cdot V_{dc} = \left[I_{dc\,\max,C} \cdot V + \sigma I_{dc\,\max,P}(V-N)\right] \cdot V_{dc} \tag{2.15}$$

where $V = \frac{v_{\text{in}}}{V_{\text{in,max}}}$, $I_{dc\,\max,C}$ is the dc current of the carrier amplifier and $I_{dc\,\max,P}$ is the dc current of the peaking amplifier at the maximum drive.

Thus, the efficiency is given by

$$\begin{aligned} \eta &= \frac{P_{RF}}{P_{dc}} \times 100 \\ &= \left[\frac{I_{1,C}^2 \cdot V^3}{(I_{1,C} + \sigma I_{1,P}) \cdot V - N \cdot \sigma I_{1,P}} + \frac{\sigma I_{1,P} \cdot (V-N) \cdot \{(I_{1,C} + \sigma I_{1,P}) \cdot V - N \cdot \sigma I_{1,P}\}}{4 I_{1,C}}\right] \\ &\div \left[I_{dc\,\max,C} \cdot V + \sigma I_{dc\,\max,P} \cdot (V-N)\right] \times 100 \end{aligned} \tag{2.16}$$

where

$$I_{dc,C} = \frac{I_{\max}}{2\pi} \cdot \frac{2\sin\frac{\alpha_c}{2} - \alpha_C \cos\left(\frac{\alpha_C}{2}\right)}{1 - \cos\left(\frac{\alpha_c}{2}\right)}, \quad I_{dc,P} = \frac{I_{\max}}{2\pi} \cdot \frac{2\sin\frac{\alpha_P}{2} - \alpha_P \cos\left(\frac{\alpha_P}{2}\right)}{1 - \cos\left(\frac{\alpha_P}{2}\right)} \tag{2.17}$$

Fig. 2.3A shows the calculated efficiencies based on the above analysis for $\alpha_C = 180°$–$360°$ and $\alpha_p = 140°$ with the uneven power drive ($\sigma = 2.3$–$2.46$). It is shown that uneven ($\sigma = 2.3$) drives with $u_c - 180^{\circ}$ and $\alpha_p = 140^{\circ}$ is the optimum operation. For the $\alpha_c$ larger than $180^{\circ}$, the efficiency drops due to the class AB operation although the Doherty amplifier is properly uneven driven. At the optimum condition, the load impedance of

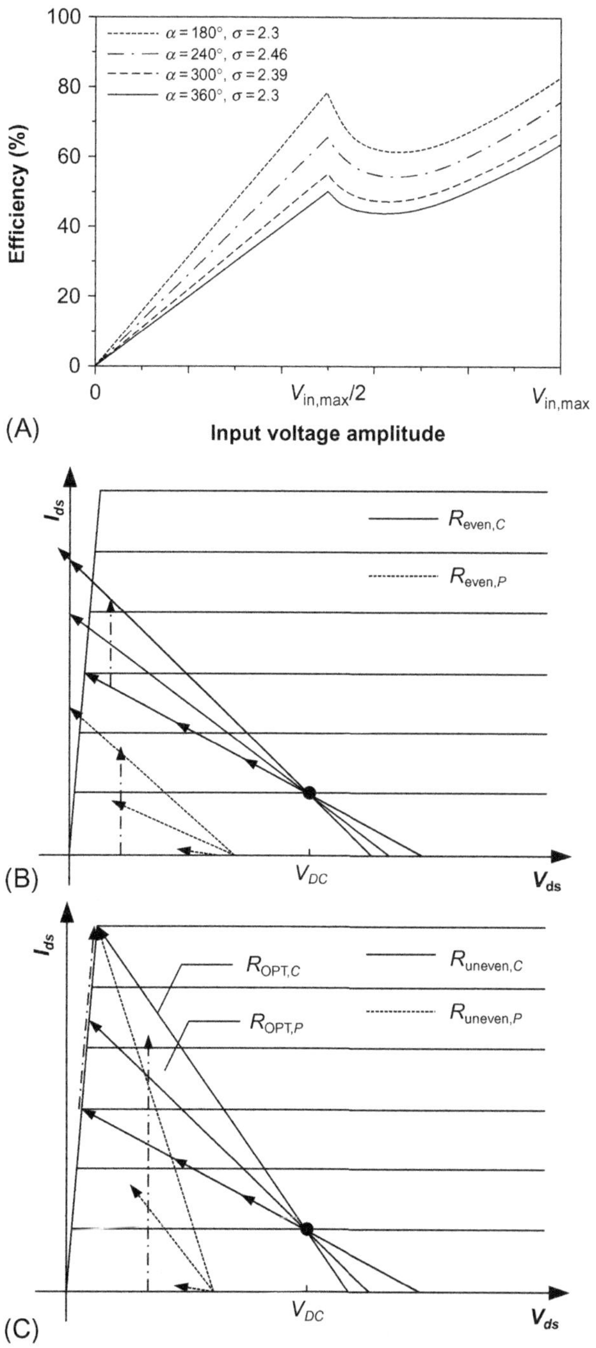


**Fig. 2.3** (A) Efficiency versus input drive level for the amplifier with uneven power drive [$\alpha_c = 180°–360°$ and $\alpha_p = 140°$ with uneven power drive ($\sigma = 2.3–2.46$)]. Load lines of the two amplifiers for (B) even power drive and (C) uneven power drive.

the carrier amplifier at the low-power region is two times larger than that of a conventional class AB amplifier.

The carrier amplifier reaches to the saturation state at the input voltage of $V_{\text{in,max}}/2$ since the maximum fundamental current swing is a half and the maximum voltage swing reaches to $V_{dc}$, as shown in Fig. 2.3B and C. As a result, the maximum power level is half of the carrier amplifier's allowable power level, and the efficiency of the amplifier is equal to the maximum efficiency of the carrier amplifier given by Eq. (2.13). Fig. 2.3B represents the load lines of the two amplifiers with the even power drive at a higher-power level. It is shown that the carrier amplifier is saturated and the voltage swing is larger than the allowable maximum voltage swing of $V_{dc}$ due to the high load impedance. The peaking amplifier at the peak power operation produces a current far less than the full capability with the voltage swing larger than $V_{dc}$. Thus, it is clear that the two amplifiers of a conventional Doherty amplifier are heavily saturated, degrading the linearity at the high-power region, and produce far less power. However, for the uneven drive, the load line of the carrier amplifier follows the knee current region without producing saturation due to proper load modulation, as shown in Fig. 2.3C. The voltage swing of the peaking amplifier increases in proportion to the power level and reaches to the maximum voltage swing of $V_{dc}$ only at the maximum power. At the maximum power level, both amplifiers have the optimum matching impedance, as shown in Fig. 2.3C. Therefore, the asymmetrical Doherty power amplifier with the uneven drive operates more linearly and produces more power than the even drive operation, delivering the expected Doherty performance.

## 2.1.2 Gate Bias Adaptation to Compensate the Low Current of Peaking Amplifier

### *2.1.2.1 Peaking Amplifier Adaptation*

The proper Doherty load modulation can be achieved by adaptively adjusting the gate-bias voltage of the peaking amplifier, increasing the bias with its input signal envelope. Fig. 2.4 shows the peaking current profile before and after the gate-bias adaption. The peaking amplifier with the higher bias voltage enhances the current to the proper level for the ideal load modulation shown in Fig. 1.3A. The gate-bias voltage increases from the pinch-off voltage at $V_{\text{in,max}}/2$ and reaches the same gate voltage of the carrier amplifier at the peak power (can be a little higher considering the class C operation). Since the input power is proportional to the square of the input voltage, the gate voltage is increased square root of the input power as shown in Fig. 2.5. The resulting Doherty amplifier has an ideal load modulation characteristic with the maximized efficiency and linearity.

Fig. 2.6 shows a block diagram of the Doherty amplifier with the gate-bias adaptation. The Doherty amplifier consists of two separate circuits, a fully matched Doherty amplifier

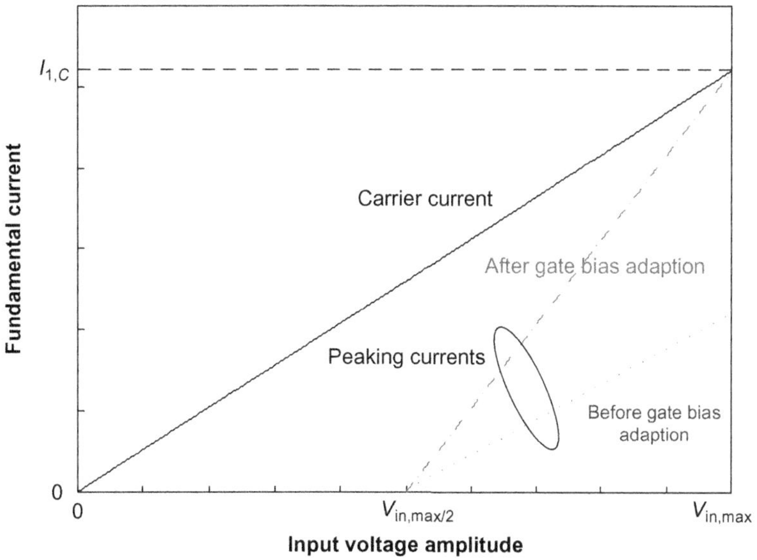


**Fig. 2.4** The current profile of peaking amplifier with the gate-bias adaptation.

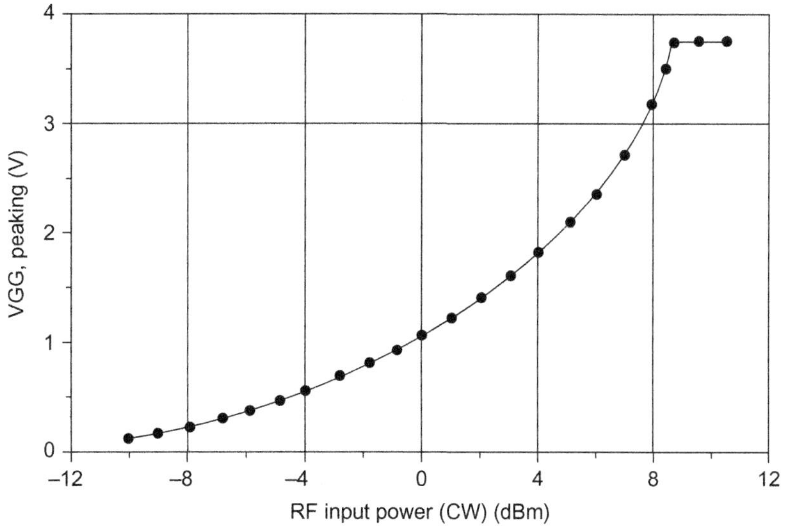


**Fig. 2.5** Gate control voltage for adaptation of the peaking amplifier.

and an adaptive gate-bias control circuit for the peaking amplifier. In real implementation, the conduction angle difference, knee voltage effect, nonlinear transconductance and capacitances distort the ideal operation. The gate-bias voltage of the carrier amplifier can be adapted for further enhanced performance in this situation.

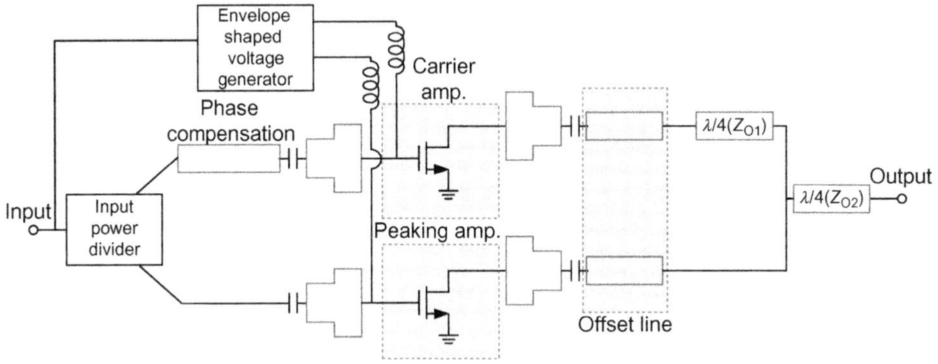


**Fig. 2.6** Schematic of the Doherty amplifier with gate-bias adaptation.

### *2.1.2.2 Adaptation of the Both Amplifiers*

In the real Doherty amplifier operation, the carrier amplifier at the low-power operation with the $2R_{OPT}$ load generates more than a half of the peak power because the effective knee voltage is reduced by the larger load resistance and the voltage swing becomes larger. This knee effect will be discussed in the following section. Fig. 2.7 shows the simulation results of the carrier amplifier using a GaN device. As depicted in Fig. 2.7A, the power level is 2.5 dB lower than that with $R_{OPT}$. Therefore, the conventional Doherty amplifier considering the knee effect has the first peak efficiency at the power level higher than the 6 dB back-off point. The gain at the $2R_{OPT}$ is lower than the expected one since the input is matched at the peak power, generating mismatch at the low-power operation.

To solve the problem, the gate bias of the carrier amplifier is increased from a deep AB bias at a low-power operation to a bias close to class A as the power level increases. The input of the carrier amplifier is conjugate-matched at the deep AB bias level for a higher gain at the back-off power level since the gain at the high power with the class A operation is higher than that at the low-power operation. The peak power of the

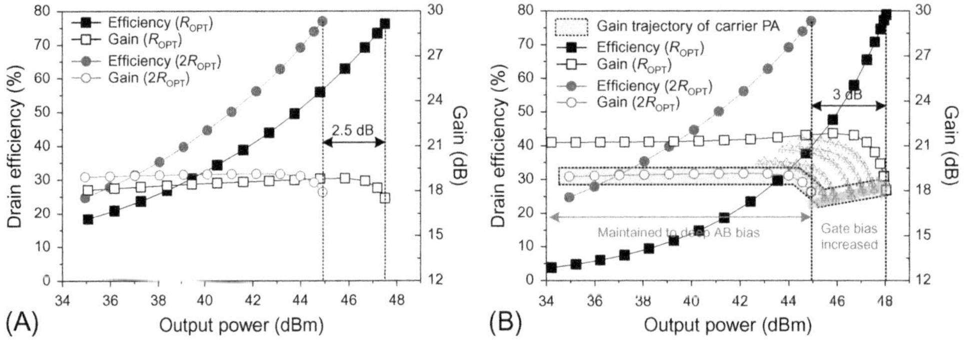


**Fig. 2.7** Simulated results for power performance of the carrier amplifier. (A) Conventional fixed class AB bias case. (B) Adapted bias case with a deep AB bias for $2R_{OPT}$ load and a class A bias for $R_{OPT}$ load.

carrier amplifier with $R_{OPT}$ is 3 dB higher than that with $2R_{OPT}$. The carrier amplifier is operated at the saturated mode in this power region, but the gain of the carrier amplifier can be maintained at a high level as indicated by the arrow line. Fig. 2.7B shows the efficiency and gain behaviors of the carrier amplifier with the gate-bias adaptation, which follows the ideal operation.

Fig. 2.8 represents the drain efficiency and gain of the peaking amplifier with the gate-bias adaptation. Before the load modulation begins, the peaking amplifier needs to be turned off, and the gate bias of the peaking amplifier should be at a C bias. As the load is modulated to $R_{OPT}$, the gate bias of the peaking is increased to an AB bias level to generate the proper current, the same level with the carrier's. The input of the peaking amplifier is conjugate-matched at the AB operation for a higher gain at the high power level. As shown in the figure, the gate bias control of the peaking amplifier is similar to the peaking control only case in the previous section.

Fig. 2.9 shows a continuous wave (CW) simulation result of the bias-controlled Doherty amplifier. The gain of the gate-bias-controlled Doherty does not decrease at the modulation region, while that of the uncontrolled Doherty amplifier decreases. As a result, the drain efficiency of the controlled Doherty amplifier is high at the 6 dB back-off power. The peak power of the Doherty amplifier is maximized by generating the full powers from the carrier and peaking amplifiers. Each gate-bias control profile is depicted in Fig. 2.10. To get the proper Doherty operation, the gate bias of the carrier amplifier is increased from deep AB bias to A bias, delivering 3 dB higher output peak power. The gate bias of the peaking amplifier is set to a C bias to prevent the turn-on. As the input power increases, the gate bias of the peaking is increased rapidly from the 6 dB back-off power point. Because drain efficiency of the Doherty amplifier is degraded at the peaking bias higher than AB bias, the gate bias of the peaking PA is increased to an AB bias level. The drain efficiency of the adapted Doherty amplifier between the two peak efficiency points is maintained at a high level.

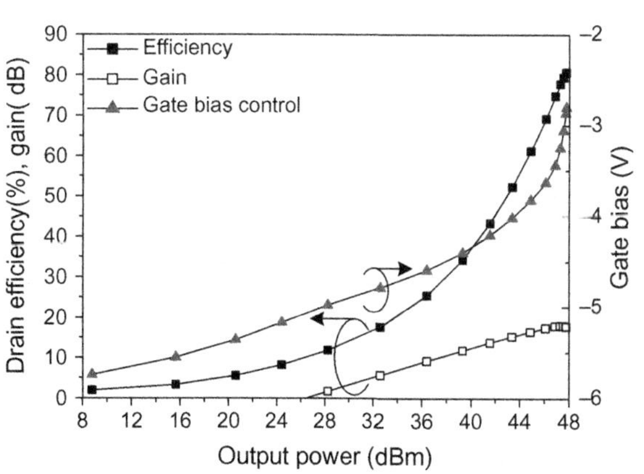


**Fig. 2.8** The performance of the gate-bias-controlled peaking amplifier.

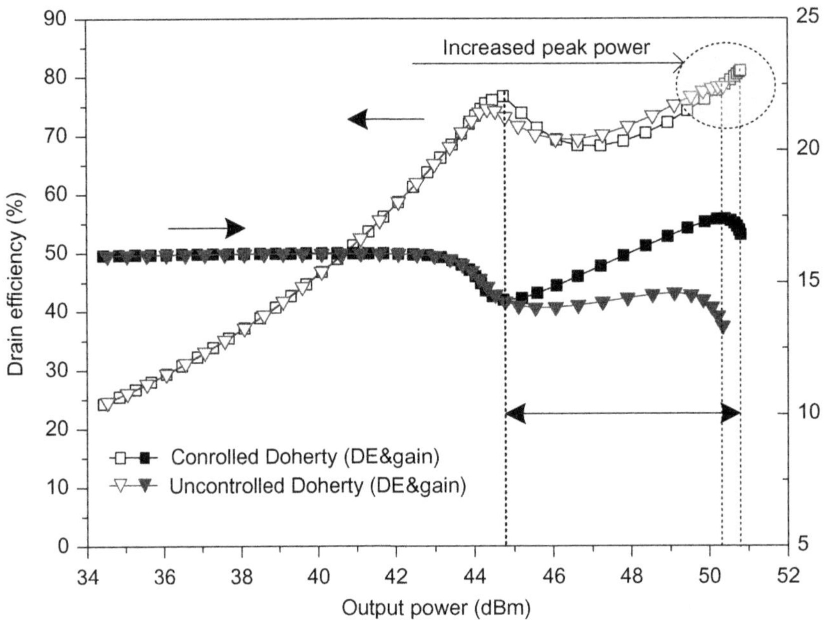


**Fig. 2.9** Gain and efficiency characteristics of the gate-bias controlled and uncontrolled Doherty amplifiers.

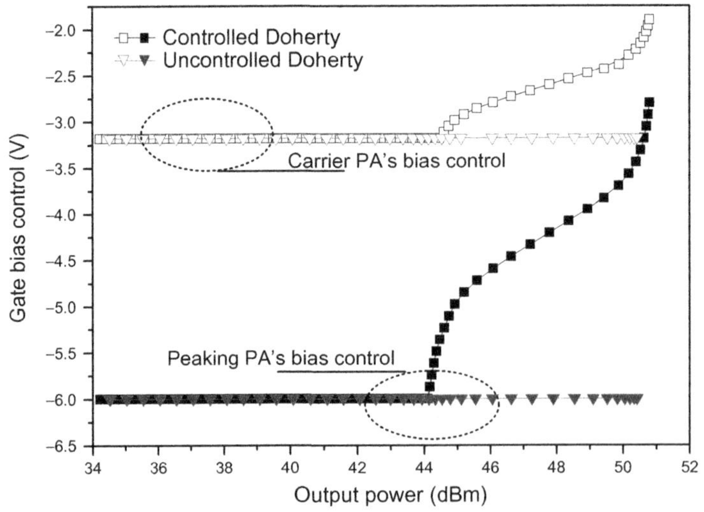


**Fig. 2.10** The controlled bias voltage shapes of the carrier and peaking amplifiers.

## 2.2 KNEE VOLTAGE EFFECT ON DOHERTY AMPLIFIER OPERATION

The knee voltage of a transistor plays an important role for the load modulation behavior of Doherty amplifier. Since the knee voltage of a transistor is smaller at a low current level, the transistor generates more power and the optimum load impedance at the 3 dB back-off power for the carrier amplifier is larger than $2R_{\mathrm{OPT}}$. Therefore, the modulated load impedance of the carrier amplifier at the back-off power, when the peaking amplifier is turned on, should be larger than $2R_{\mathrm{OPT}}$. Otherwise, the first peak power region is located at less than 6 dB back-off power point. Because of the larger load modulation range required due to the knee voltage effect, the load modulation circuit should be modified accordingly.

### 2.2.1 Doherty Amplifier Operation With Knee Voltage

Theoretically, the Doherty amplifier has the highest efficiency at the 6 dB back-off and the peak power levels. The maximum efficiencies at the two power points can be achieved through the perfect load modulation of the carrier and peaking amplifiers. The perfect load modulation of the carrier amplifier with a zero knee voltage, depicted in Fig. 2.11A, indicates that the output powers for both $R_{\mathrm{OPT}}^{\mathrm{Ideal}}$ and $2R_{\mathrm{OPT}}^{\mathrm{Ideal}}$ are $P_1^{\mathrm{Ideal}}$ and $P_1^{\mathrm{Ideal}}/2$, respectively. The efficiencies ($\eta^{\mathrm{Ideal}}$'s) are also the same at the two power levels; the carrier amplifier is in equally saturated states for the both cases. In this operation, $P_1^{\mathrm{Ideal}}$, $\eta^{\mathrm{Ideal}}$, and $R_{\mathrm{OPT}}^{\mathrm{Ideal}}$ are given by

$$P_1^{\mathrm{Ideal}} = \frac{1}{2} \cdot I_{1,C}(\theta_C) \cdot V_1 \tag{2.18}$$

$$\eta^{\mathrm{Ideal}} = \frac{P_1^{\mathrm{Ideal}}}{P_{dc}^{\mathrm{Ideal}}} \times 100 = \frac{(1/2) \cdot I_{1,C}(\theta_c)}{I_{dc,C}(\theta_c)} \times 100 \tag{2.19}$$

$$R_{\mathrm{OPT}}^{\mathrm{Ideal}} = \frac{V_1}{I_{1,c}(\theta_c)} \tag{2.20}$$

where

$$I_{1,C}(\theta_c) = \frac{I_{\max}}{2\pi} \cdot \frac{\theta_c - \sin\theta_c}{1 - \cos(\theta_c/2)} \tag{2.21}$$

$$I_{dc,C}(\theta_c) = \frac{I_{\max}}{2\pi} \cdot \frac{2 \cdot \sin(\theta_c/2) - \theta_c \cos(\theta_c/2)}{1 - \cos(\theta_c/2)} \tag{2.22}$$

$I_{1,C}(\theta_C)$ and $I_{dc,C}(\theta_C)$ are the fundamental and dc current components, respectively, of the carrier amplifier biased at the conduction angle $\theta_c$. In the ideal case, $V_1$ is equal to $V_{dc}$ because of the zero on-resistance with zero knee voltage.

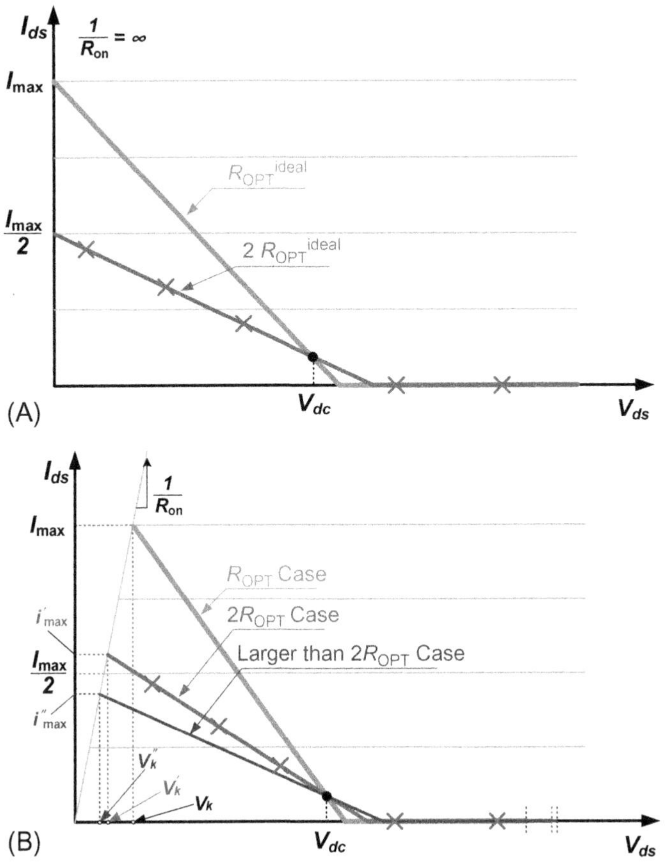


**Fig. 2.11** Load line of the carrier amplifier: (A) ideal case with zero knee voltage and (B) practical case with nonzero knee voltage.

In a real device, the effect of the knee voltage should be considered. Fig. 2.11B shows a load line with the knee voltage $V_k$. For $R_{\mathrm{OPT}}$ and $2R_{\mathrm{OPT}}$, the load impedances are given by

$$R_{\mathrm{OPT}} = \frac{V_{dc} - V_k}{I_{1,c}(\theta c)} \tag{2.23}$$

$$R^{\mathrm{Case\_I}} = 2R_{\mathrm{OPT}} = \frac{V_{dc} - V_k'}{i_{1,c}'(\theta_c)} \tag{2.24}$$

where

$$i_{1,c}'(\theta_c) = \frac{i_{max}'}{2\pi} \cdot \frac{\theta_c - \sin\theta_c}{1 - \cos(\theta_c/2)} \tag{2.25}$$

$$V_k = R_{\text{on}} \cdot I_{\max}$$

$$V_k' = R_{\text{on}} \cdot i'_{\max}$$

In Case I with $2R_{\text{OPT}}$, $i'_{\max}$ can be derived from the loadlines in Fig. 2.11

$$i'_{\max} = \frac{I_{\max} V_{dc}}{2V_{dc} - I_{\max} R_{\text{on}}} \tag{2.26}$$

$i'_{\max}$ represents the maximum current when the carrier amplifier has the load impedance of $2R_{\text{OPT}}$. Thus, the output powers for the $R_{\text{OPT}}$ and $R^{\text{Case\_I}}$ cases can be calculated as follows:

$$P_{R_{\text{OPT}}} = \left(\frac{1}{2}\right) \cdot I_{1,c}(\theta_c) \cdot (V_{dc} - V_k) \tag{2.27}$$

$$P^{\text{Case\_I}} = \left(\frac{1}{2}\right) \cdot i'_{1,c}(\theta_c) \cdot (V_{dc} - V_k') \tag{2.28}$$

The powers are plotted in Fig. 2.12A. For the calculation, it is assumed that $V_{dc}$ is 30 V and $I_{\max}$ is 8 A with uniform transconductance. $\theta_c$ is fixed at 210°. As shown, the carrier amplifier with $2R_{\text{OPT}}$ delivers the power more than a half of $P_{R_{\text{OPT}}}$ because of the enlarged voltage and current swings from $(V_{dc} - V_k)$ to $(V_{dc} - V_k')$ and from $I_{\max}/2$ to $i'_{\max}$, respectively. The efficiency at the 6 dB back-off region does not reach to the peak value because the carrier amplifier, which is optimally designed for $R_{\text{OPT}}$ load, is not in the saturation state when the peaking amplifier is turned on.

To maximize the efficiency at the 6 dB back-off power region, the carrier amplifier with nonzero knee voltage should have a load impedance larger than $2R_{\text{OPT}}$, like Case II in Fig. 2.11B. As shown, the current is reduced, but the voltage swing is increased, maintaining the 3 dB lower power. In this case, the output power can be written as

$$P^{\text{Case\_II}} = \left(\frac{1}{2}\right) \cdot i''_{1,c}(\theta_c) \cdot (V_{dc} - V_k'') \tag{2.29}$$

where

$$i''_{1,c}(\theta_c) = \frac{i''_{\max}}{2\pi} \cdot \frac{\theta_c - \sin\theta_c}{1 - \cos\left(\frac{\theta_c}{2}\right)} \tag{2.30}$$

$$V_k'' = R_{\text{on}} \cdot i''_{\max}$$

Since $P^{\text{Case-II}}$ should be half of $P_{R_{\text{OPT}}}$, $i''_{\max}$, the maximum current when the carrier amplifier has a load impedance larger than $2R_{\text{OPT}}$, can be calculated:

$$i''_{\max} = \frac{1}{R_{\text{on}}} \cdot \frac{V_{dc} - \sqrt{(V_{dc})^2 - 2 \cdot V_k \cdot (V_{dc} - V_k)}}{2} \tag{2.31}$$

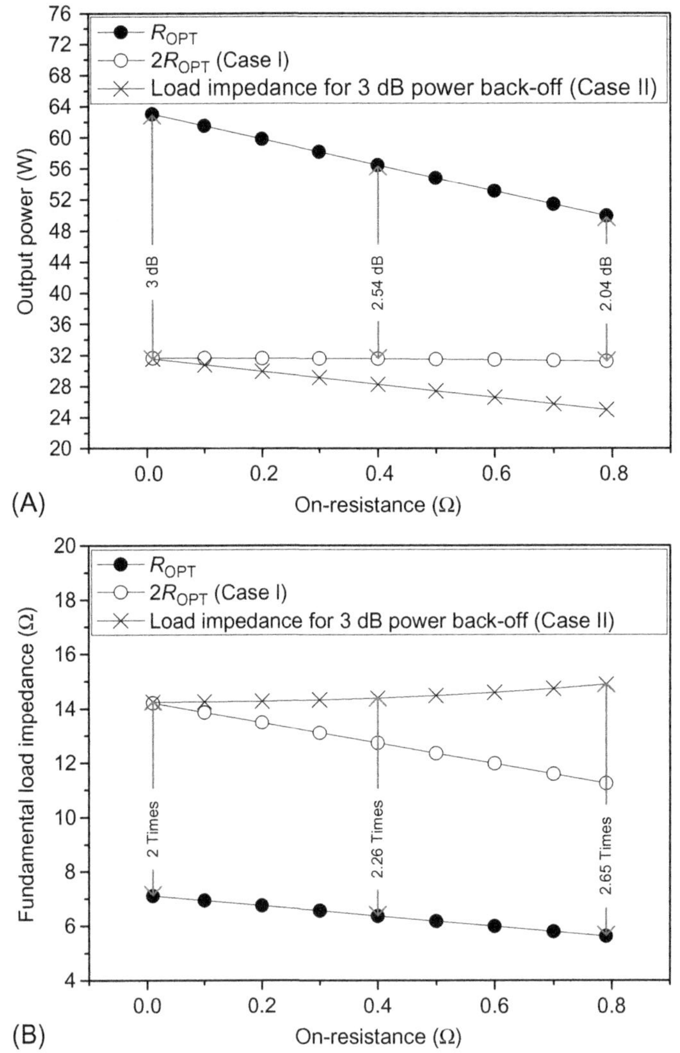


**Fig. 2.12** (A) Maximum output powers employing $R_{OPT}$, $R^{Case\text{-}I}$, and $R^{Case\text{-}II}$ versus $R_{on}$. (B) Load impedances for $R_{OPT}$, $R^{Case\text{-}I}$, and $R^{Case\text{-}II}$ versus $R_{on}$.

As shown in Fig. 2.12A, the maximum output power is decreased with $R_{on}$. But $P^{Case\_I}$ does not change much since $i'_{max}$ is increased and the knee voltage is reduced. Therefore, the difference from $P_{R_{OPT}}$ is smaller than 3 dB. Fig. 2.12B shows the required load impedance for various $R_{on}$. As $R_{on}$ increases, the load impedance of the carrier amplifier in the low-power region also increases, larger than $2R_{OPT}$.

To verify the knee voltage effects on the Doherty operation, the carrier amplifier is designed and tested using a Cree GaN HEMTCGH40045 device with a 45 W power at

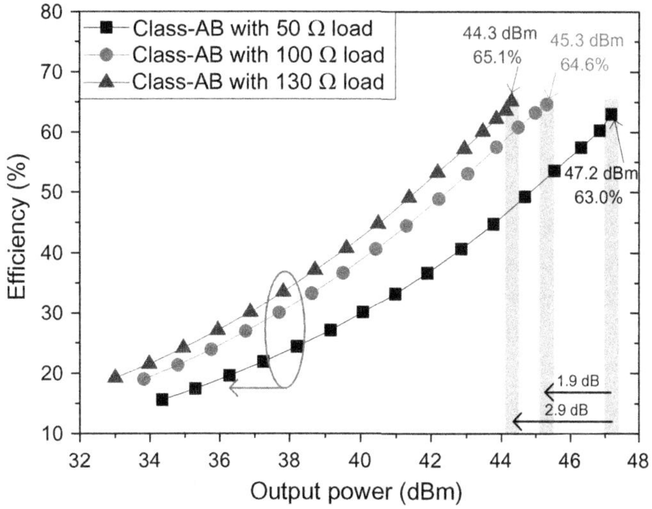


**Fig. 2.13** Load modulation results for carrier amplifier employing the load impedances of 50, 100, and 130 Ω.

2.655 GHz. First, the output of the amplifier is matched to 50 Ω. Its load impedance is then modulated by adjusting the offset line and output termination impedance. Fig. 2.13 shows the load modulation results for the carrier amplifier employing the output termination impedances of 50, 100, and 130 Ω. As expected, the PA with 100 Ω delivers maximum efficiency at 45.3 dBm, which is 1.9 dB back-off power from the maximum output power of 47.2 dBm with $R_{\mathrm{OPT}}$ of 50 Ω. On the other hand, the PA with 130 Ω has its maximum efficiency at the 2.9 dB back-off power.

## 2.2.2 Load Modulation Behavior of Doherty Amplifier With Optimized Carrier Amplifier

To properly design a Doherty amplifier using a real device, the knee voltage effect should be considered as described in Section 2.2.1. Fig. 2.14 shows the operational diagram of the Doherty amplifier used to analyze the load modulation behavior. Compared with the

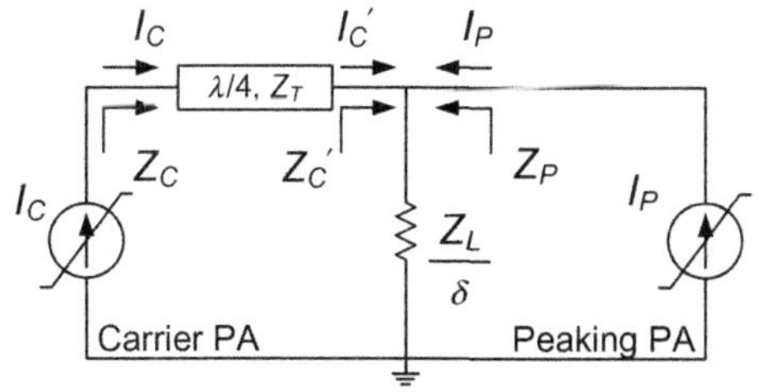


**Fig. 2.14** Operational diagram of the Doherty amplifier with knee effect.

conventional one, it has a scaled load impedance of $Z_L/\delta$. $\delta$ represents the increased resistance ratio to achieve the fully saturated operation of the carrier amplifier at the 6 dB back-off region. For the analysis, it is assumed that each current source is linearly proportional to the input voltage with the maximum current of $I_{\max}$. The ideal harmonic short circuits are provided so that the output power and efficiency can be determined by only the fundamental and dc components.

The load impedances of the carrier and peaking amplifiers are determined by their current ratio through the load modulation. The peaking amplifiers for the both cases of the with/without knee effect can generate the proper output current. However, since the load impedance at the 6 dB back-off power region of the carrier amplifier with the knee effect is larger than that of the amplifier without the knee effect, the current of the carrier amplifier cannot reach to the maximum value by the load modulation, leading to the degradation of output power. For the proper current generation, the peaking amplifier should generate more current at the peak power, by $\gamma$ times, than the carrier amplifier.

The fundamental currents, as a function of the input voltage, of the conventional Doherty amplifier and the Doherty amplifier with the knee effect, are represented in Fig. 2.15, The currents, $I_C$, $I_P$, $I_{DC,C}$ and $I_{DC,P}$ which represent the fundamental and dc currents of the carrier and peaking amplifiers with the knee effect, respectively, are derived as

$$I_C = \frac{I_{1,C}(\theta_c)}{V_{\text{in,max}}} \cdot v_{\text{in}}, \quad 0 \le v_{\text{in}} \le \frac{\gamma\beta + 1}{2\gamma} \cdot V_{\text{in,max}} \tag{2.32}$$

$$I_{DC,C} = \frac{I_{dc,C}(\theta_c)}{V_{\text{in,max}}} \cdot v_{\text{in}}, \quad 0 \le v_{\text{in}} \le \frac{\gamma\beta + 1}{2\gamma} \cdot V_{\text{in,max}} \tag{2.33}$$

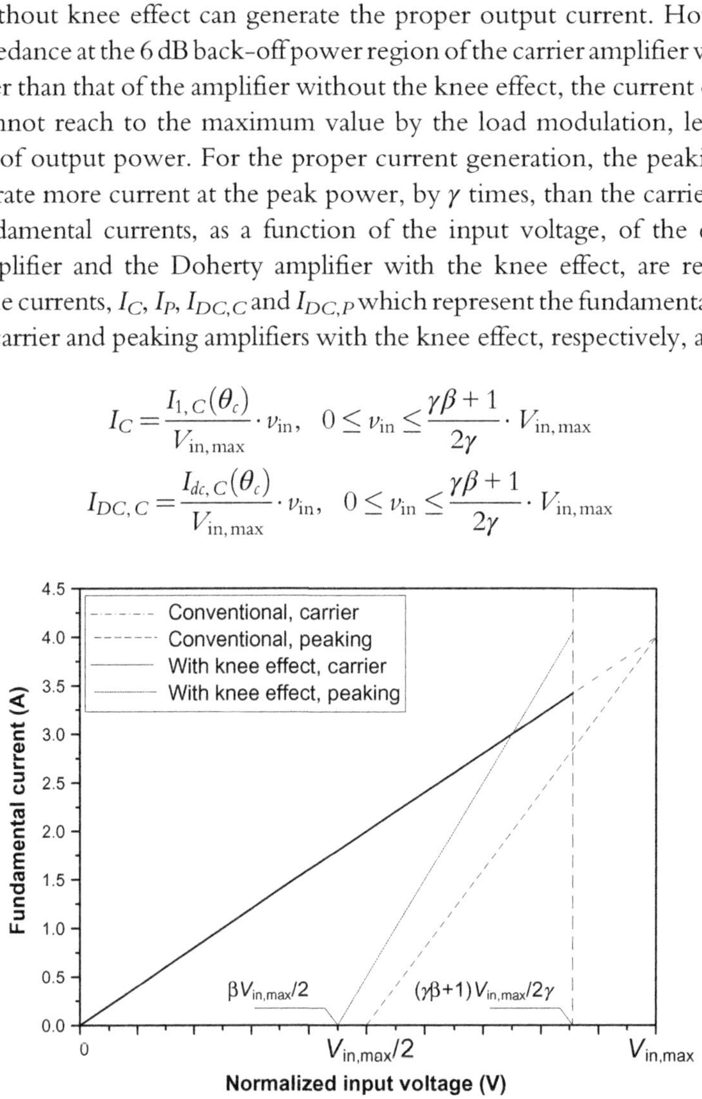


**Fig. 2.15** Fundamental currents of the conventional and knee effect Doherty amplifiers according to the input voltage ($\theta_c = 210°$ and $\theta_p = 150°$).

$$I_P = \begin{cases} 0, & 0 \le v_{\text{in}} \le (\beta/2)\, V_{\text{in,max}} \\ \gamma \cdot \sigma \cdot I_{1,p}(\theta_p) \cdot \left( \dfrac{v_{\text{in}}}{V_{\text{in,max}}} - \dfrac{\beta}{2} \right), & \left(\dfrac{\beta}{2}\right) V_{\text{in,max}} \le v_{in} \le (\gamma\beta + 1)\, V_{\text{in,max}}/2\gamma \end{cases} \tag{2.34}$$

$$I_{DC,P} = \begin{cases} 0, & 0 \le v_{\text{in}} \le \beta \cdot V_{\text{in,max}}/2 \\ \gamma \cdot \sigma \cdot I_{dc,p}(\theta_p) \cdot \left( \dfrac{v_{\text{in}}}{V_{\text{in,max}}} - \dfrac{\beta}{2} \right), & \left(\dfrac{\beta}{2}\right) V_{\text{in,max}} \le v_{\text{in}} \le (\gamma\beta + 1)\, V_{\text{in,max}}/2\gamma \end{cases} \tag{2.35}$$

Here, $I_{1,C}$ and $I_{dc,C}$ are the fundamental and dc currents of the carrier amplifier with the conduction angle of $\theta_c$, which are defined in (2.21) and (2.22).

$$I_{1,p}(\theta_p) = \frac{I_{\text{max}}}{2\pi} \cdot \frac{\theta_p - \sin\theta_p}{1 - \cos(\theta_p/2)} \tag{2.36}$$

$$I_{dc,p}(\theta_p) = \frac{I_{\text{max}}}{2\pi} \cdot \frac{2 \cdot \sin(\theta_p/2) - \theta_p \cdot \cos(\theta_p/2)}{1 - \cos(\theta_p/2)} \tag{2.37}$$

$$\beta = \frac{2 \cdot i''_{\text{max}}}{I_{\text{max}}} = \frac{V_{dc} - \sqrt{(V_{dc})^2 - 2 \cdot V_k \cdot (V_{dc} - V_k)}}{V_k} \tag{2.38}$$

$$\sigma = \frac{2 \cdot I_{1,c}(\theta_c)}{I_{1,p}(\theta_p)} \tag{2.39}$$

As explained in Section 2.2.1, the carrier amplifier with the knee effect has a load impedance at the 6 dB back-off power point which is larger than $2R_{\text{OPT}}$, and the voltage swing of the amplifier is increased, as shown in Fig. 2.11B. To generate the first peak efficiency at the 6 dB back-off level, the current of the knee effect Doherty amplifier should be lower than that of the conventional Doherty amplifier by $\beta$ as described previously. Due to the lower current, the input power at the 6 dB back-off is lower, and the input power for the peak power generation is also reduced as shown in Fig. 2.15. To get the proper modulation, the peaking amplifier should be overdriven by a factor of $\gamma$, which will be derived later. $\sigma$ indicates the uneven power input drive ratio between the carrier and peaking amplifiers.

Using the currents of the carrier and peaking amplifiers, the load impedances of the two amplifiers can be calculated:

$$Z_C = \begin{cases} \dfrac{\delta \cdot Z_T^2}{Z_L}, & 0 \le v_{\text{in}} \le \dfrac{\beta}{2} \cdot V_{\text{in,max}} \\ \dfrac{\delta \cdot Z_T^2}{Z_L \cdot \left[1 + \dfrac{I_P}{I'_C}\right]}, & \dfrac{\beta}{2} \cdot V_{\text{in,max}} \le v_{\text{in}} \le \dfrac{\gamma\beta + 1}{2\gamma} V_{\text{in,max}} \end{cases} \tag{2.40}$$

$$Z_P = \begin{cases} \infty, & 0 \le v_{\text{in}} \le \frac{\beta}{2} \cdot V_{\text{in,max}} \\ \frac{Z_L}{\delta} \cdot \left(1 + \frac{I'_C}{I_P}\right), & \frac{\beta}{2} \cdot V_{\text{in,max}} \le v_{\text{in}} \le \frac{\gamma\beta + 1}{2\gamma} V_{\text{in,max}} \end{cases} \tag{2.41}$$

where

$$\delta = \frac{R^{\text{Case\_II}}}{R^{\text{Case\_I}}} = \frac{V_{dc} + \sqrt{(V_{dc})^2 - 2 \cdot V_k \cdot (V_{dc} - V_k)}}{2 \cdot \beta \cdot (V_{dc} - V_k)} \tag{2.42}$$

Using the lossless quarter-wave transmission line in Fig. 2.14, $I'_C$ is calculated:

$$I'_C = \delta \cdot I_C \cdot \frac{Z_T}{Z_L} - I_P \tag{2.43}$$

Since the carrier amplifier of the knee effect Doherty amplifier has a larger load impedance than the conventional Doherty, the carrier amplifier operates in the heavily saturated region after the peaking amplifier is turned on. To prevent the saturated operation of the carrier amplifier, the load impedance of the carrier amplifier should be lowered by generating more current from the peaking amplifier. Thus, the peaking amplifier should be overdriven by a factor of $\gamma$. For linear operation of the carrier amplifier without voltage clipping, the voltage across the current source of the carrier amplifier should satisfy the following condition at the maximum input voltage:

$$Z_C \cdot I_C \le V_{dc} - \frac{\gamma\beta + 1}{2\gamma} \cdot V_k \tag{2.44}$$

From Eq. (2.44), $\gamma$ can be inferred:

$$\gamma \ge \frac{2\delta R_{\text{OPT}} I_{1,C}(\theta_c) + V_k}{2V_{dc} - 2R_{\text{OPT}} I_{1,C}(\theta_c)(\beta\delta - 1) - \beta V_k} \tag{2.45}$$

In this analysis, it is assumed that $Z_T = R_{\text{OPT}}$ and $Z_L = R_{\text{OPT}}/2$. $\delta$ represents the $R^{\text{Case\_I}}$ to $R^{\text{Case\_II}}$ ratio to achieve the fully saturated operation of the carrier amplifier at the back-off region. For the conventional case, $\delta$, $\gamma$, and $\beta$ are equal to 1. Thus, the fundamental load impedance of the carrier amplifier is modulated from $2R_{\text{OPT}}$ to $R_{\text{OPT}}$, while that of the peaking amplifier varies from $\infty$ to $R_{\text{OPT}}$. On the other hand, for the knee effect Doherty amplifier, $\delta$ and $\gamma$ are $>1$, but $\beta$ is $<1$. Thus, the fundamental load impedances of the carrier and peaking amplifiers are modulated from $2\delta R_{\text{OPT}}$ to $2R_{\text{OPT}}\ \{\delta(\gamma\beta + 1) - \gamma\}/(\gamma\beta + 1)$ and from $\infty$ to $R_{\text{OPT}}\ (\gamma\beta + 1)/2\gamma$, respectively, as described in Fig. 2.16A. To explore the load modulation behavior, $V_k$ of 4 V is assumed, and $R_{\text{OPT}}$ is calculated using Eq. (2.20). At the low-power region ($0 \le v_{\text{in}} \le \beta V_{\text{in,max}}/2$),

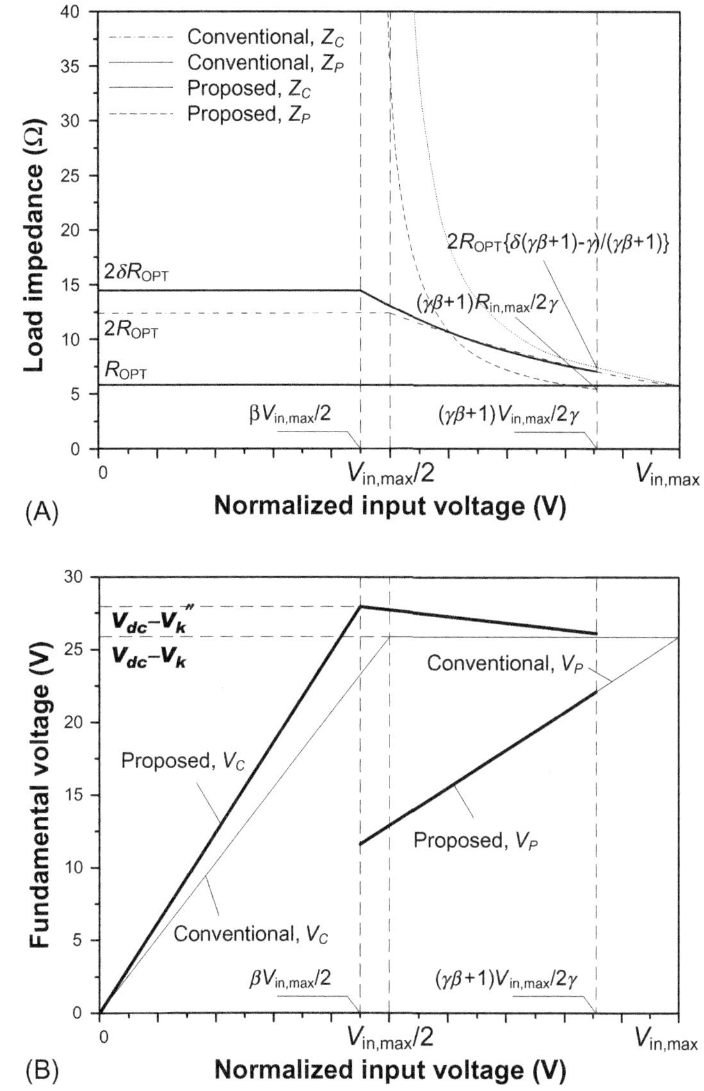


**Fig. 2.16** (A) Load impedances of the conventional and proposed Doherty amplifier according to the input voltage. (B) Fundamental voltages of the carrier ($V_c$) and peaking ($V_p$) amplifiers for the conventional and knee effect Doherty amplifiers.

the load impedance of the carrier amplifier is larger than $2R_{\rm OPT}$, that is, $2\delta R_{\rm OPT}$. At the high-power region ($\beta V_{\rm in,max}/2 \leq v_{in} \leq (\gamma\beta+1)V_{\rm in,max}/2\gamma$), the load impedance maintains a value larger than conventional amplifier. In contrast, the load impedance of the peaking amplifier sustains a smaller value than that of the conventional amplifier; this results in a slight power degradation.

Fig. 2.16B shows the resulting fundamental voltages of the carrier and peaking amplifiers for the conventional and knee effect Doherty amplifiers. As expected, the carrier amplifier with the knee effect has a larger fundamental voltage at the 6 dB back-off because of the larger load impedance. Therefore, the carrier amplifier can be fully saturated, delivering the high efficiency.

Using the currents and load impedances of the carrier and peaking amplifiers, the efficiency is estimated through MATLAB. Fig. 2.17 represents the calculated efficiencies based on the above analysis for the $\theta_c = 210°$ and $\theta_p = 150°$ operation. To calculate the efficiencies, $\delta$, $\beta$, $\sigma$, and $\gamma$ are determined based on $V_{dc}$, $V_k$, $I_{1,c}(\theta_c)$, and $I_{1,p}(\theta_p)$, which are 30 V, 4 V, 4.2129 A, and 3.6384 A, respectively. The resulting $\delta$, $\beta$, $\sigma$, and $\gamma$ are 1.1725, 0.9235, 2.3158, and 1.2494, respectively. In the low-power region, the knee effect Doherty amplifier has higher efficiency than that of the conventional amplifier because of the larger load impedance. In particular, the knee effect Doherty amplifier delivers its maximum efficiency of 67.9% at the 6 dB back-off output power, which is an increase of about 5% compared with the conventional one. Although the maximum output power and efficiency at the maximum output power are slightly degraded because of the imperfect load modulation, the knee effect Doherty scheme improves efficiency for amplification of a modulated signal without any additional circuitry. Since the currents and voltages of the carrier and peaking amplifiers are different, the impedances are mismatched at the peak power. This mismatch problem can be solved by the offset line control, which will be described in Section 2.3.

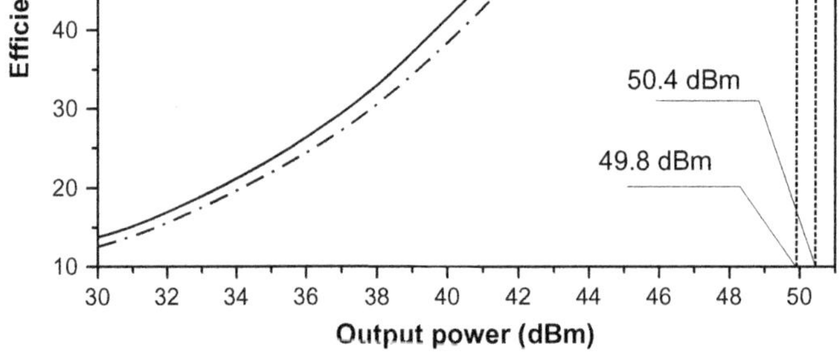

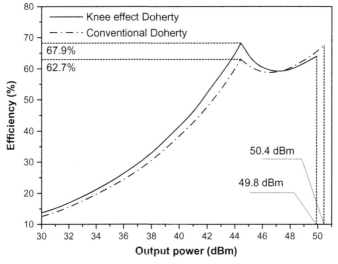

**Fig. 2.17** Efficiencies of the conventional and knee effect Doherty amplifier according to the output power.

## 2.3 OFFSET LINE DESIGN FOR COMPENSATION OF PEAKING AMPLIFIER PHASE VARIATION

In the Doherty operation, the peaking amplifier is biased at class C mode. With the bias, the input and output capacitances of the device vary a lot during the turn-on transition, and the amplifier has a poor AM-PM characteristic. This phase variation of the peaking amplifier produces the phase mismatch between the carrier and peaking amplifier paths, disturbing the load modulation of the Doherty amplifier. In a conventional Doherty amplifier design, the phases of the peaking and carrier amplifiers are matched at the maximum power level, and the mismatch is significant when the peaking amplifier operates at a low output power. However, the two peak efficiencies of the Doherty operation are maintained. The offset line cannot compensate this variable capacitance effect, but the phase variation effect can be minimized by proper design of the offset lines.

### 2.3.1 Phase Variation of the Peaking Amplifier

To analyze the load modulation behavior of the Doherty amplifier, the amplifier model is simplified as shown in Fig. 2.18. The carrier and peaking currents are normalized to the peak current and have variable phases. Because the offset line compensates the output capacitance effects of the device, the amplifier with the offset line can be substituted by an ideal current source with an impedance level of $R_O$.

Fig. 2.19A shows the simulated results of the load modulations for the carrier and peaking amplifiers, without the additional offset line, as the output power of the amplifier is increased. Without the phase variation of the peaking amplifier, the output loads of the carrier and peaking amplifiers are ideally modulated from 100 to 50 Ω and infinite to 50 Ω, respectively, following the resistive load. However, the peaking amplifier has a negative phase variation as shown in Fig. 2.19B, which is extracted from Cree CGH40045 GaN device model. With the phase variation, the output loads of the carrier and peaking amplifiers are properly matched at the two maximum efficiency points, and

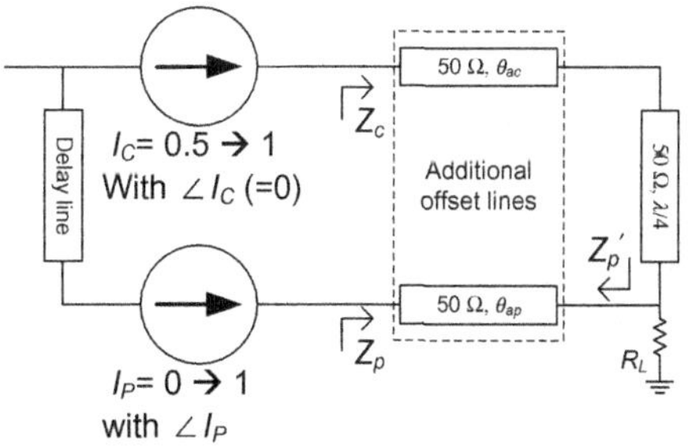


**Fig. 2.18** Simplified Doherty amplifier model for simulation of the load modulation by the offset lines.

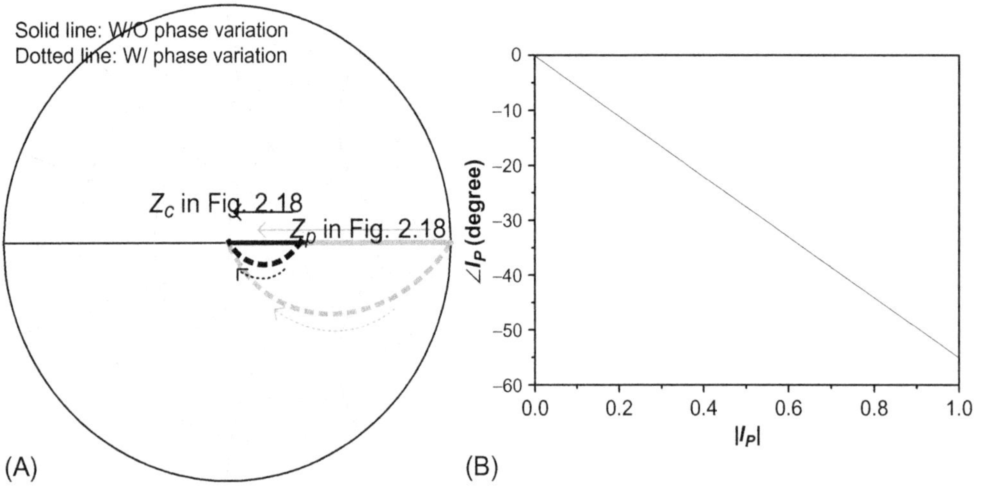


**Fig. 2.19** (A) Load modulations of the carrier and peaking amplifiers with/without the phase variation of the peaking amplifier. (B) Phase variation of the peaking amplifier (the reference point is at the class B bias).

the load cannot be properly modulated between the two points as shown in Fig. 2.19A. The improper load modulation degrades the efficiency of the Doherty amplifier.

This load mismatch problem can be solved somewhat by inserting additional offset lines at the carrier and peaking amplifiers. In addition, the new offset lines increase the load impedances of the carrier and peaking amplifiers, mitigating the knee effect of the carrier amplifier as described in Section 2.2 and enhancing the efficiency of the Doherty amplifier.

## 2.3.2 Load Modulation of Peaking Amplifier With the Additional Offset Lines

To solve the phase variation problem of the peaking amplifier, additional offset lines having delays of $\theta_{ac}$ and $\theta_{ap}$ are added at the outputs of the current sources as shown in Fig. 2.18. In this Doherty amplifier, the relationship between the normalized carrier current $I_c$ and normalized peaking current $I_p$ are given by

$$I_c = \frac{1+I_p}{2} e^{j\left(\frac{1}{4}\pi + \theta_{ac} + \theta_{ap}\right)} \tag{2.46}$$

With only the additional peaking offset line ($\theta_{ac}=0$), the impedances $Z_p$ and $Z_c$, defined in Fig. 2.18, are expressed in terms of $I_p$ as

$$Z_p = \frac{Z_0^2}{2R_L}\left(\left(1+\frac{1}{I_p}\right) + j\tan\theta_{ap}\left(1-\frac{1}{I_p}\right)\right)$$

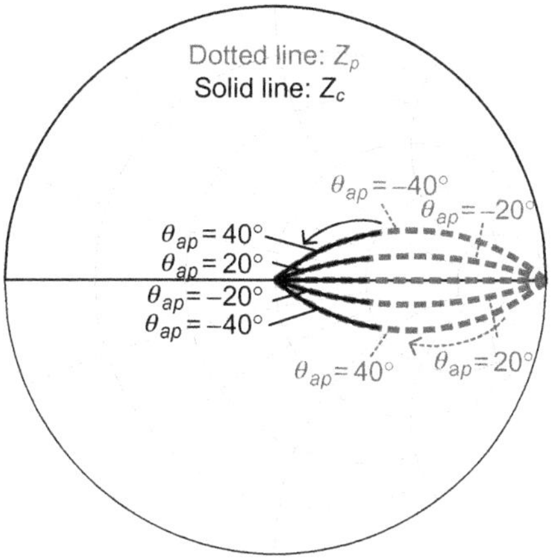


**Fig. 2.20** Load modulation of Doherty amplifier according to the additional peaking offset line ($\theta_{ac} = 0°$) without the phase variation of the peaking amplifier.

$$Z_c = Z_0\left(\frac{Z_0}{R_L} - \frac{2I_p}{1+I_p} + j\tan\theta_{ap}\left(1 - \frac{2I_p}{1+I_p}\right)\right) \tag{2.47}$$

Fig. 2.20 shows the load modulation curves calculated by Eq. (2.47) for the case with the additional offset line but without the peaking amplifier phase variation. The phase variation of the modulated loads with a negative length of the additional offset line is opposite to the phase variation due to the peaking amplifier phase variation shown in Fig. 2.19, indicating that the phase variation of the peaking amplifier can be compensated by shorting the conventional offset line, but at the expense of the carrier matching. We will also address the carrier amplifier problem in the next section.

If there is a phase difference between the carrier and peaking amplifiers ($\theta_d$), $Z_p$ in Eq. (2.47) is expressed by

$$Z_p = \frac{1+I_p}{2I_p}\left(\cos\theta_d - \tan\theta_{ap}\sin\theta_d\right) + jZ_0\left(\tan\theta_{ap} - \frac{1+I_p}{2I_p}\left(\sin\theta_d + \cos\theta_d\tan\theta_{ap}\right)\right) \tag{2.48}$$

To minimize the imaginary term of the $Z_p$, $\theta_{ap}$ should be

$$\theta_{ap} = \tan^{-1}\left(\frac{\frac{1+I_p}{2I_p}\sin\theta_d}{1 - \frac{1+I_p}{2I_p}\cos\theta_d}\right) \tag{2.49}$$

Because the phase of the peaking amplifier linearly decreases from 0 to $\theta_{d,\max}$ with the $I_p$ as shown in Fig. 2.19B, Eq. (2.49) can be rearranged to

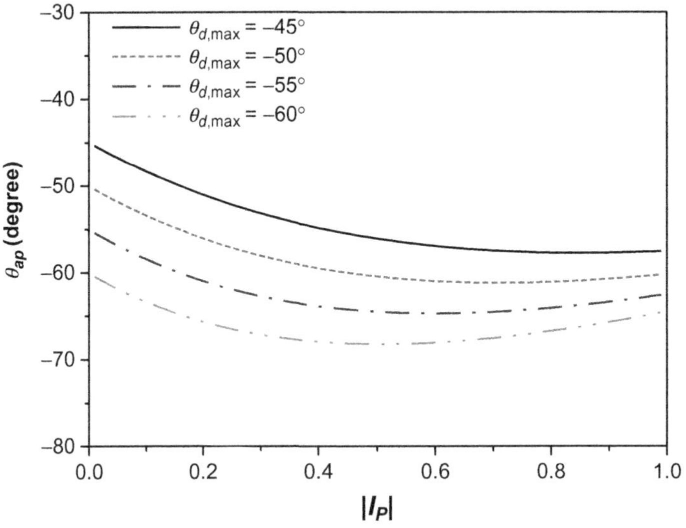


**Fig. 2.21** Calculated additional peaking offset line lengths according to the magnitudes of $I_P$ and $\theta_{ap}$.

$$\theta_{ap} = \tan^{-1}\left(\frac{\frac{1+I_p}{2I_p}\sin\left\{\left(I_p-1\right)\theta_{d,\max}\right\}}{1-\frac{1+I_p}{2I_p}\cos\left\{\left(I_p-1\right)\theta_{d,\max}\right\}}\right) \tag{2.50}$$

The calculated additional offset line lengths with the linear phase variation of the peaking amplifier with $\theta_{d,\max}$ of −60 to −45° are shown in Fig. 2.21. Although the required additional offset line length is different according to the $I_p$, the difference is not much. Therefore, it is possible to compensate the phase variation of the peaking amplifier by adjusting the length of the offset line.

With the calculated additional offset line length required at the $I_p$ of 0.5, the load modulation prolifes of the carrier and peaking amplifier are shown in Fig. 2.22. As analyzed by Eq. (2.50), the load modulation of the peaking amplifier with the additional peaking offset line is near to the ideal Doherty load modulation.

However, due to the additional peaking offset line with the negative length, the impedance toward the peaking amplifier (at $Z_p'$ in Fig. 2.18) is not an open but an inductive impedance. This impedance affects the load modulation of the carrier amplifier as shown in Fig. 2.22.

## 2.3.3 The Load Modulation of the Carrier Amplifier With the Additional Offset Lines

For the proper load modulation of the carrier amplifier with the additional peaking offset line, the phase angle of $Z_c$ when the peaking amplifier is in the off- state should be minimized. To convert the carrier load to a resistive load, an additional offset line is needed at the output of the carrier amplifier also. Fig. 2.23 shows the load modulation of the Doherty amplifier

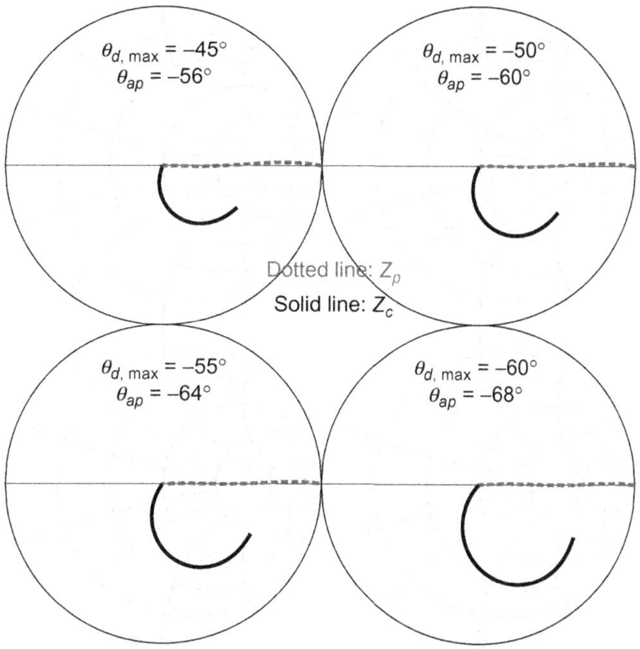


**Fig. 2.22** The calculated load modulations of the carrier and peaking amplifiers with the additional peaking offset lines.

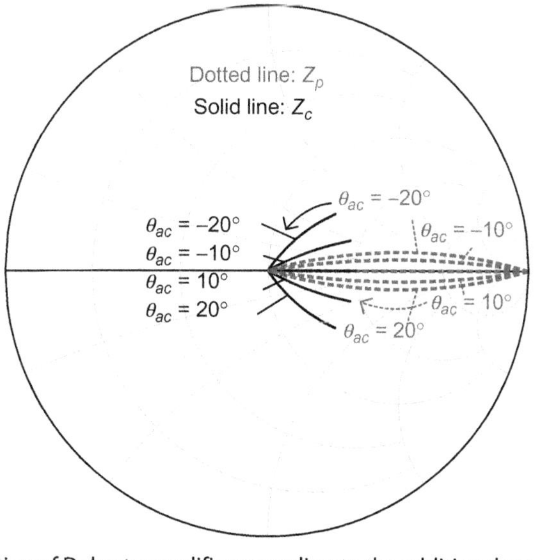


**Fig. 2.23** Load modulation of Doherty amplifier according to the additional carrier offset line ($\theta_{ap} = 0°$) without considering the phase variation of the peaking amplifier.

according to the additional carrier offset line but without the phase variation of the peaking amplifier. With the negative length of additional offset line (shorten the conventional offset line), $Z_c$ at the low-power region becomes an inductive impedance, which can absorb the impedance variation generated by the compensation of the peaking amplifier phase variation. With the negative additional carrier offset line, the load modulation of the peaking amplifier becomes slightly inductive. Therefore, if the negative additional offset line is added at the carrier amplifier, the additional peaking offset line should be a little lengthened.

Fig. 2.24A shows the load modulation of the Doherty amplifier with the additional offset lines. To compensate the phase variation of the peaking amplifier (Fig. 2.19B), the additional offset lines for the carrier and peaking amplifiers are set to $-43°$ and $-13°$, respectively. The load of the peaking amplifier is modulated similarly to the ideal case. The load of the carrier amplifier does not closely follow the ideal modulation, but the phase variation is small, and the impedance deviation is also small. The magnitude of $Z_c$ at the low-power region is larger than that with the conventional offset line.

With the larger load for the carrier amplifier, the efficiency at the low-power region is increased. Moreover, the load modulation is ideal for the Doherty amplifier with the knee voltage effect as discussed in Section 2.2. As discussed before, the load of the carrier amplifier could not be properly matched at $50\,\Omega$ due to the larger load modulation range required for the carrier amplifier. However, with these offset lines, both the larger load of the carrier amplifier at the low power region and the proper power match at the peak output power can be realized simultaneously as shown in Fig. 2.24B.

In addition, the $Z_p$ is resistive with the larger value than the ideal Doherty modulation. Due to the larger load of the peaking amplifier during the load modulation, the peaking amplifier can get a high efficiency, even higher than that of the ideal Doherty amplifier. In Fig. 2.24B, $Z_c$ of the Doherty amplifier with the additional offset lines starts to modulate from the $-5$ dB output power back-off level, not $-6$ dB, because the output power delivered from the ideal current source is proportional to the magnitude of the load. However, the output power delivered from a real transistor is decreased for the larger load, and the $Z_c$ is modulated at the output power back-off level larger than 5dB as shown in the following section.

## 2.3.4 Simulation Results With Real Device

To validate the investigations in the previous sections, we have designed a symmetrical Doherty amplifier (Fig. 2.25). The Doherty amplifier is designed at the 1.94 GHz using GaN pHEMT devices (Cree CGH40045) with the drain bias of 50 V. The gate of the peaking amplifier is biased at $-6$ V, and the phase of the amplifier is varied from 0° to $-50°$. For the conventional offset line design, the offset line lengths of the carrier and peaking amplifiers are 50°. For the new offset line design, the offset line lengths of the carrier and peaking amplifiers should be 37° ($-13°$ of $\theta_{ac}$) and 10° ($-40°$ of $\theta_{ap}$), respectively. For

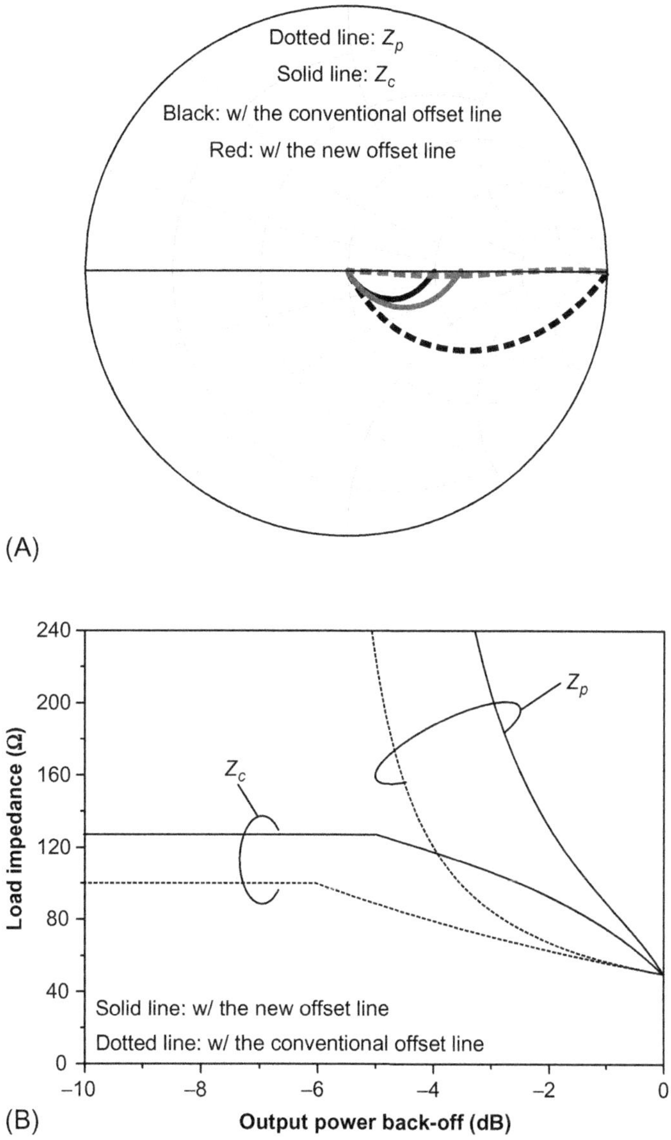


**Fig. 2.24** Load modulation of Doherty amplifier with the additional offset lines (A) on the Smith chart and (B) on the magnitude graph according to the output power back-off level.

comparison, the Doherty amplifier without the phase variation of the peaking amplifier is also designed. This amplifier is identical to the adaptive phase control to compensate the phase variation of the peaking amplifier using the dual input Doherty structure.

Figs. 2.26 and 2.27 show the load modulation behaviors of the carrier and peaking amplifiers under the three different operation conditions. When the conventional offset

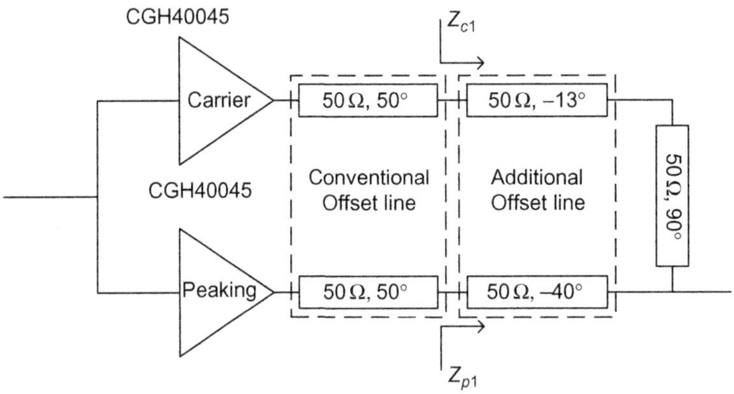


**Fig. 2.25** Schematic Doherty amplifier for the simulation.

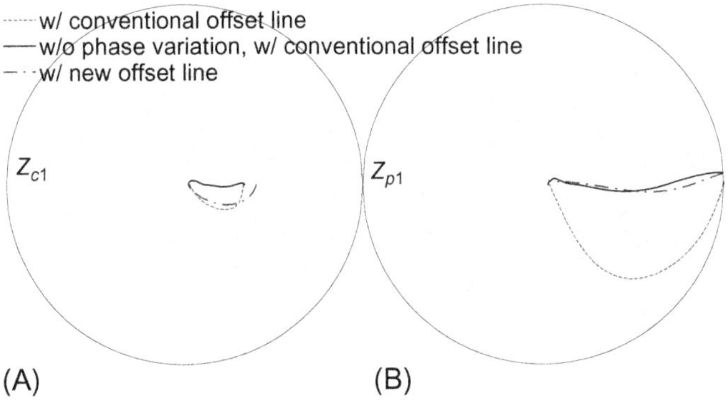


**Fig. 2.26** Load modulations of the simulated Doherty amplifier: (A) carrier amplifier and (B) peaking amplifier.

lines are added, the phase variation of the peaking amplifier disturbs the load modulation. The improper modulation can be compensated either by the additional offset lines or by the adaptive phase control at the dual input structure (equivalent to the without peaking phase variation and conventional offset lines). The load modulation is examined for the reference planes defined by $Z_{c1}$ and $Z_{p1}$ shown in Fig. 2.25 for the three cases with/without the peaking amplifier phase variation and the new offset line with the phase variation. Following the expectation explained at Section 2.3.3, $Z_{p1}$ with the new offset is modulated following the resistive load with a larger magnitude. $Z_{c1}$ with the new offset line still has an imaginary component during the load modulation, but the magnitude of the $Z_{c1}$ at the low-power region is larger than that with the conventional offset line. The load modulation is started at −6 dB back-off power level.

Fig. 2.28 shows the efficiencies of the carrier and peaking amplifiers. Due to the larger $Z_{c1}$ with the new offset line, the efficiency of the carrier amplifier is higher at the

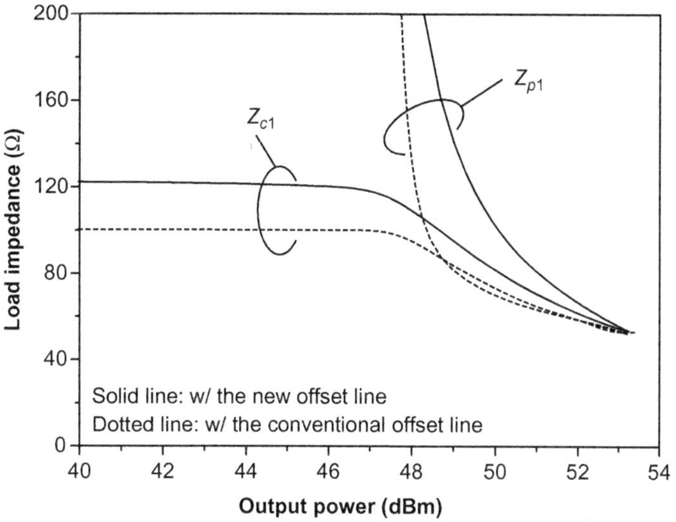


**Fig. 2.27** Magnitudes of simulated $Z_{c1}$ and $Z_{p1}$ during the load modulation.

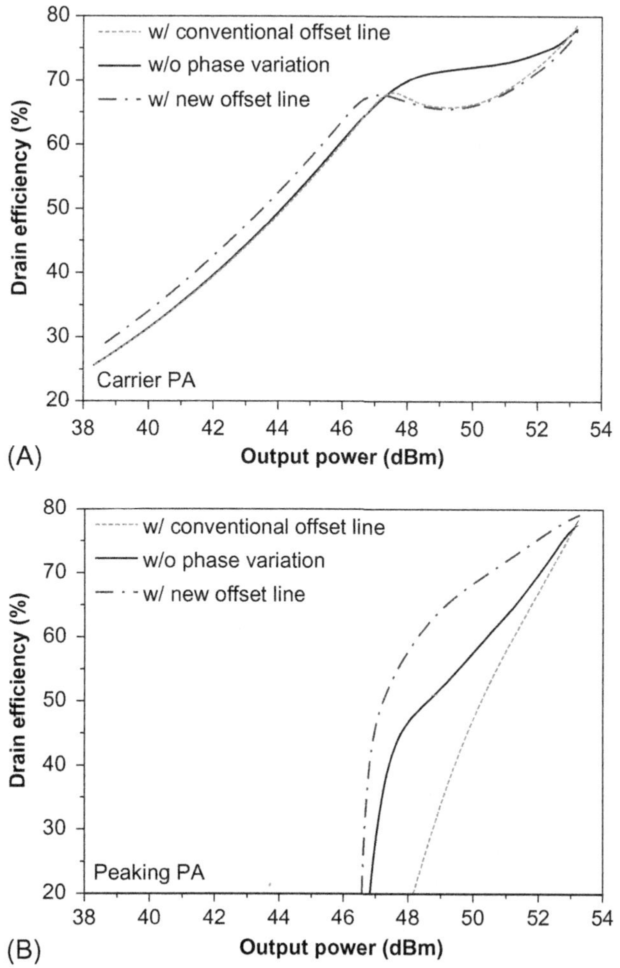


**Fig. 2.28** Simulated drain efficiencies of the (A) carrier amplifier and (B) peaking amplifier.

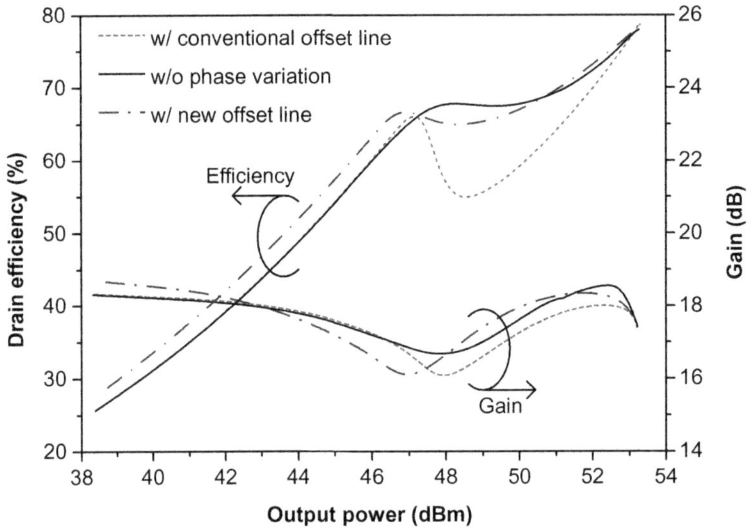


**Fig. 2.29** Simulated drain efficiencies and gains of the Doherty amplifier.

low-power region than those of the other operation conditions. However, the imaginary component of the load still exists, and the efficiency of the carrier amplifier at the high-power region is a little lower than that without the phase variation and similar to that with the conventional offset line. The efficiency of the peaking amplifier with the new offset line is higher than those with the other cases at the all output power region since the magnitude of the $Z_{p1}$ is larger than those of the other cases.

Fig. 2.29 shows the simulated CW performance of the Doherty amplifier. With the new offset line, the efficiency of the Doherty amplifier is higher at the low-power region. At the high-power region, the efficiency of the carrier amplifier is lower than that without the phase variation. However, the efficiency of the peaking amplifier is higher, and the efficiency of the Doherty amplifier at the high-power region is similar to that without the phase variation. Therefore, the Doherty amplifier with the new offset line not only provides a higher efficiency than that with the conventional offset line but also has a higher efficiency than that without the phase variation of the peaking amplifier. Gain characteristic does not show much difference for the three cases. However, due to the larger load operation of the carrier amplifier, the gain of the Doherty amplifier with the new offset line shows early compression characteristic.

## FURTHER READING

[1] Y. Yang, et al., Optimum design for linearity and efficiency of microwave Doherty amplifier using a new load matching technique, Microw. J. 44 (12) (2001) 20–36.
[2] B. Kim, et al., The Doherty power amplifier, IEEE Microw. Mag. 7 (5) (2006) 42–50.
[3] J. Kim, et al., Optimum operation of asymmetrical-cells-based linear Doherty power amplifiers-uneven power drive and power matching, IEEE Trans. Microwave Theory Tech. 53 (5) (2005) 1802–1809.

[4] Y. Yang, et al., A microwave Doherty amplifier employing envelope tracking technique for high efficiency and linearity, IEEE Microwave Wireless Compon. Lett. 13 (9) (2003) 370–372.
[5] Y. Park, et al., Gate bias adaptation of Doherty power amplifier for high efficiency and high power, IEEE Microwave Wireless Compon. Lett. 25 (2) (2015) 136–138.
[6] J. Moon, et al., Efficiency enhancement of Doherty amplifier by mitigating the knee voltage effect, IEEE Trans. Microwave Theory Tech. 59 (1) (2011) 143–152.
[7] R. Quaglia, et al., Offset lines in Doherty power amplifiers: analytical demonstration and design, IEEE Microwave Wireless Compon. Lett. 23 (2) (2013) 93–95.
[8] S. Kim, et al., Accurate offset line design of Doherty amplifier with compensation of peaking amplifier phase variation, IEEE Trans. Microwave Theory Tech. 64 (10) (2016) 3224–3231.
[9] R. Darraljі, et al., A dual-input digitally driven Doherty amplifier architecture for performance enhancement of Doherty transmitters, IEEE Trans. Microwave Theory Tech. 59 (5) (2011) 1284–1293.
[10] P. Colantonio, et al., Increasing Doherty amplifier average efficiency exploiting device knee voltage behavior, IEEE Trans. Microwave Theory Tech. 59 (9) (2011) 2295–2305.

CHAPTER THREE

# Enhancement of Doherty Amplifier

Various techniques to enhance the performance of the Doherty amplifier are introduced. The drain bias of the peaking amplifier is applied at a higher voltage than that of the carrier. It can control the first peak efficiency back-off level, and the knee effect can be absorbed. The optimized carrier and peaking amplifier designs are introduced to maximize the power performance. The Doherty amplifier can be designed using saturated amplifiers for carrier and peaking amplifiers to maximize the efficiency. These optimized design methods are discussed. Finally, the average power tracking operation technique is also introduced.

## 3.1 DOHERTY AMPLIFIER WITH ASYMMETRIC $V_{DS}$

In the previous chapter, we have analyzed the Doherty amplifier with the optimized carrier amplifier considering the knee voltage effect. It is shown that, to get the maximum efficiency at the 6 dB back-off output region, the carrier amplifier should have a load modulation range larger than two, contrary to the ratio of two for the ideal Doherty amplifier.

As another approach for the Doherty amplifier design considering the knee voltage effect, an asymmetrical drain bias of the carrier and peaking amplifiers can be considered. To get the maximum efficiency at the 6 dB back-off region, the carrier amplifier should generate smaller power with $2R_{OPT}$ load, and it is operated at the drain bias lower than that of the peaking amplifier. The drain bias of the carrier amplifier is adjusted to get the 3 dB output power variation by larger load modulation using a bigger size peaking amplifier, and the bias voltage is determined by the on-resistance ($R_{on}$) of the amplifier. In this design, the Doherty amplifier can deliver the first peak efficiency at the 6 dB back-off output power region. The same analysis can be applied to have the first peak efficiency at lower or higher than the 6 dB back-off point by adjusting the drain bias ratio between the carrier and peaking amplifiers.

Under the assumption that the amplifiers are biased at class B mode as before, the operation is analyzed using the load lines of the amplifiers shown in Figs. 3.1 and 3.2. Fig. 3.1 shows the load line of carrier amplifier at the back-off output power region, where only the carrier amplifier is operated. The $R_{on}$ is determined by

$$R_{on} = \frac{V_k}{I_{max}} = \frac{V_{k1}}{I_{max2}/2} = \frac{V_{k2}}{I_{max2}} \tag{3.1}$$

*Doherty Power Amplifiers*
https://doi.org/10.1016/B978-0-12-809867-7.00003-X

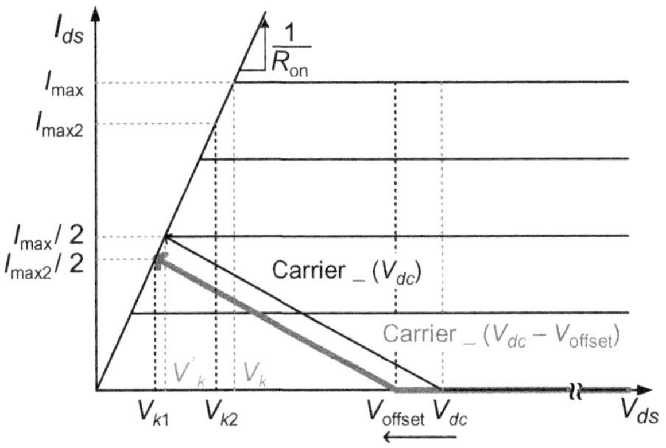


**Fig. 3.1** Load lines of the carrier amplifier at the back-off output power state.

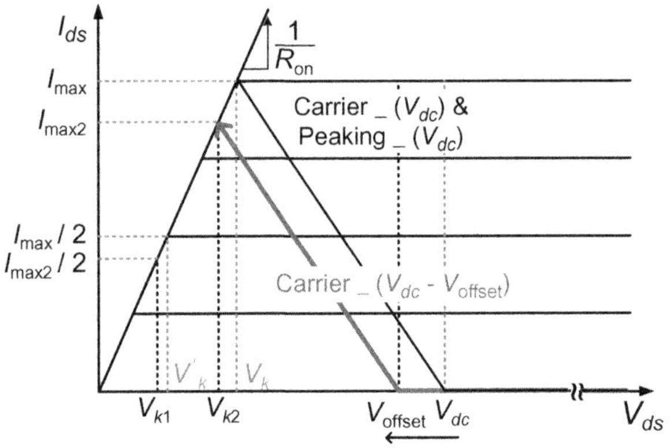


**Fig. 3.2** Load lines of the carrier and peaking amplifiers at the peak output power state.

The symbols are defined in the figure. The knee voltages at the first peak efficiency point before and after changing the drain bias voltage of the carrier amplifier are assigned to $V'_k$ and $V_{k1}$, respectively. $I_{max}$ and $I_{max2}$ are the maximum currents for the load lines, respectively. The $V_{offset}$ is the reduced amount of the drain bias voltage from $V_{dc}$ due to the knee effect. The fundamental load impedance ($R_{OPT}$), output power of the carrier amplifier ($P_{BO}$), and efficiency at the first peak efficiency region ($\eta_{BO}$) can be derived as

$$R_{OPT} = \frac{V_{dc} - V_{offset} - V_{k1}}{0.5 \cdot 0.5 \cdot I_{max2}} = \frac{4 \cdot V_{dc} - 4 \cdot V_{offset} - 2 \cdot I_{max2} \cdot R_{on}}{I_{max2}} \tag{3.2}$$

$$\begin{aligned} P_{BO} &= \frac{1}{2} \cdot (V_{dc} - V_{\text{offset}} - V_{k1}) \cdot \frac{1}{2} \cdot \frac{1}{2} \cdot I_{\max 2} \\ &= \frac{1}{8} \cdot I_{\max 2} \cdot \left( V_{dc} - V_{\text{offset}} - \frac{I_{\max 2}}{2} \cdot R_{\text{on}} \right) \end{aligned} \tag{3.3}$$

$$\eta_{BO} = \frac{\frac{1}{8} \cdot I_{\max 2} \cdot \left( V_{dc} - V_{\text{offset}} - \frac{I_{\max 2}}{2} \cdot R_{\text{on}} \right)}{(V_{dc} - V_{\text{offset}}) \cdot (1/\pi) \cdot 0.5 \cdot I_{\max 2}} = \frac{\pi}{4} \cdot \left[ 1 - \frac{I_{\max 2} \cdot R_{\text{on}}}{2 \cdot (V_{dc} - V_{\text{offset}})} \right] \tag{3.4}$$

Fig. 3.2 presents the load line of the carrier and peaking amplifiers at the peak output power, where both amplifiers generate their full power. The fundamental load impedance, output power, and efficiency of the carrier amplifier at the region can be derived as

$$R_{\text{OPT,PEP\_Carrier}} = \frac{V_{dc} - V_{\text{offset}} - I_{\max 2} \cdot R_{\text{on}}}{0.5 \cdot I_{\max 2}} \tag{3.5}$$

$$P_{\text{PEP\_Carrier}} = \frac{1}{4} \cdot I_{\max 2} \cdot (V_{dc} - V_{\text{offset}} - I_{\max 2} \cdot R_{\text{on}}) \tag{3.6}$$

$$\eta_{\text{PEP\_Carrier}} = \frac{\pi}{4} \cdot \left( 1 - \frac{I_{\max 2} \cdot R_{\text{on}}}{V_{\text{dc}} - V_{\text{offset}}} \right) \tag{3.7}$$

With the similar calculation for the peaking amplifier, the overall output power and the efficiency of the Doherty amplifier at the peak power can be represented by following equations:

$$\begin{aligned} P_{\text{PEP}} &= \frac{1}{4} \cdot I_{\max} \cdot (V_{dc} - V_k) + \frac{1}{4} \cdot I_{\max 2} \cdot (V_{dc} - V_{\text{offset}} - V_{k2}) \\ &= \frac{1}{4} \cdot I_{\max} \cdot \left[ V_{dc} - V_k + \frac{V_{k2}}{V_k} \cdot (V_{dc} - V_{\text{offset}} - V_{k2}) \right] \\ &= \frac{1}{4} \cdot I_{\max} \cdot \left[ \left( 1 + \frac{I_{\max 2}}{I_{\max}} \right) \cdot V_{dc} - \frac{I_{\max 2}}{I_{\max}} \cdot V_{\text{offset}} - R_{\text{on}} \cdot \left( I_{\max} + \frac{I_{\max 2}^2}{I_{\max}} \right) \right] \end{aligned} \tag{3.8}$$

$$\begin{aligned} \eta_{\text{PEP}} &= \frac{\text{PEP}}{V_{dc} \cdot \frac{I_{\max}}{\pi} + (V_{dc} - V_{\text{offset}}) \cdot \frac{I_{\max 2}}{\pi}} \\ &= \frac{\pi}{4} \cdot \frac{\left[ \left( 1 + \frac{I_{\max 2}}{I_{\max}} \right) \cdot V_{dc} - \frac{I_{\max 2}}{I_{\max}} \cdot V_{\text{offset}} - R_{\text{on}} \cdot \left( I_{\max} + \frac{I_{\max 2}^2}{I_{\max}} \right) \right]}{\left( 1 + \frac{\text{I}_{\max 2}}{\text{I}_{\max}} \right) \cdot V_{dc} - \frac{I_{\max 2}}{I_{\max}} \cdot V_{\text{offset}}} \end{aligned} \tag{3.9}$$

Thus, the output back-off level in dB, $P_{\text{back-off}}$ (dB), can be obtained from Eqs. (3.3), (3.8):

$$P_{\text{back-off}}\,(\text{dB}) = 10 \cdot \log(P_{\text{PEP}}) - 10 \cdot \log(P_{B_o}) \tag{3.10}$$

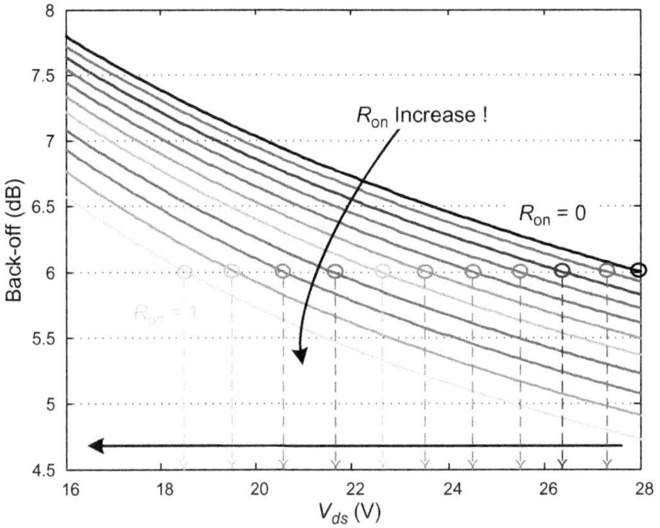


**Fig. 3.3** Back-off power level of Doherty amplifier versus $R_{on}$ (0.1 Ω step) for different $V_{dc}$ of carrier amplifier. The drain bias voltage of peaking amplifier is 28 V.

The back-off level in Eq. (3.10) is calculated using MATLAB simulation as a function of $R_{on}$ and is depicted in Fig. 3.3. The original $V_{dc}$ of the Doherty amplifier is assumed to be 28 V. Under the load modulation condition from $2R_{OPT}$ to $R_{OPT}$, $V_{dc}$ of the carrier amplifier, which delivers the 6 dB back-off level, is decreased as $R_{on}$ is increased. If the carrier amplifier is biased at the $V_{dc}$, the back-off level is reduced, lower than 6 dB for all $R_{on}$. This means that the Doherty amplifier does not reach to the maximum efficiency at the 6 dB back-off output power level but at a higher power level. Therefore, the peaking amplifier should be turned on at the higher output power level in order to have the maximum efficiency at the first peak point. In this operation, since the peaking amplifier is turned on at the higher output power level, the load of the carrier amplifier cannot be modulated properly, and the carrier amplifier gets into the saturated region. However, this simulation result shows that with the reduced carrier amplifier's $V_{dc}$, the peak efficiency can be achieved at the 6 dB back-off level. This curve also shows that, with the further reduced drain bias of the carrier amplifier, the back-off power point can be adjusted to a lower power region if it is needed to amplify the signal with a large peak-to-average power (PAPR).

Fig. 3.4A shows the efficiencies of the Doherty amplifier at the back-off output power (BO) and peak output power (PEP) level when the 6 dB back-off level is maintained by reducing the drain bias voltage of the carrier amplifier. As the $R_{on}$ is increased, the fundamental output voltage swing is reduced, and the efficiencies at both output power levels are decreased. Because the knee voltage is increased from $V_{k1}$ to $V_{k2}$ due to the

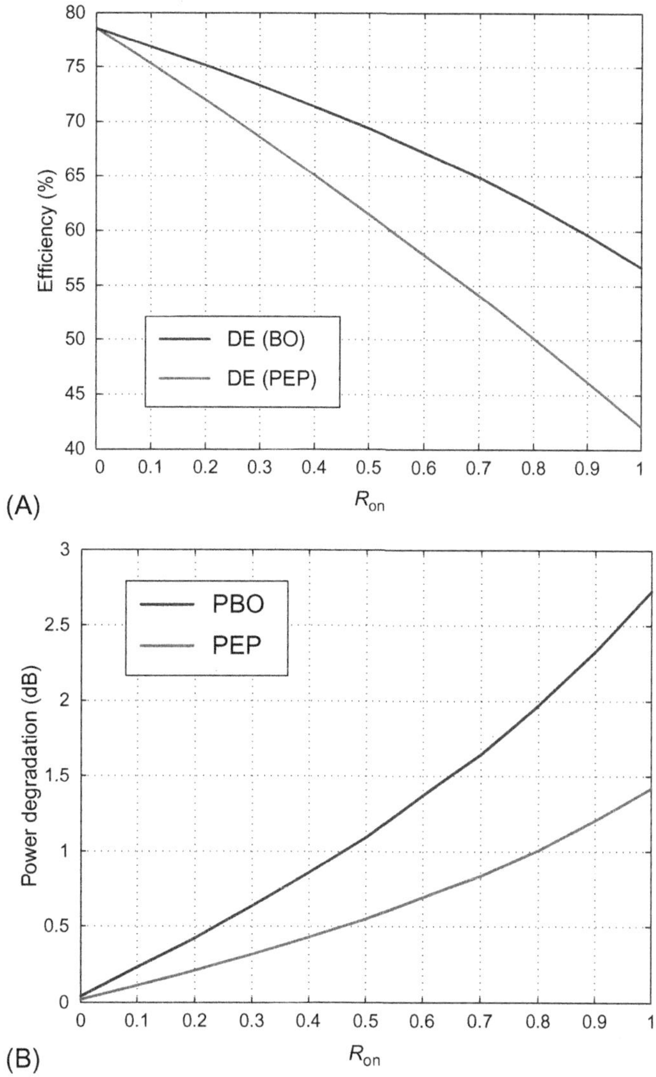


**Fig. 3.4** (A) Efficiency degradation of the Doherty amplifier versus $R_{on}$ and (B) output power degradation of the Doherty amplifier versus $R_{on}$.

load modulation, the efficiency at the peak output power is lower than that at the back-off output power level. Fig. 3.4B shows the output power degradation when the 6 dB back-off level is maintained. Because the $V_{dc}$ of the carrier amplifier is decreased further with a larger $R_{on}$, the fundamental output current and voltage of the carrier amplifier are further reduced, resulting in the output power degradation. However, the output power degradation at the peak output power is smaller than that at the back-off output power level due to the peaking amplifier power generation with $V_{dc}$.

Based on the analysis, a Doherty amplifier has been designed using GaN HEMT device. The $V_{dc}$ of the peaking amplifier is 28 V and that of the carrier amplifier is reduced to 22.4 V-based $R_{on}$ of 0.65 Ω in Fig. 3.3. Fig. 3.5A and B shows the efficiencies of the conventional Doherty and the modified Doherty amplifier. By reducing the $V_{dc}$ of the carrier amplifier to 22.4 V, the maximum efficiency could be achieved accurately at the 6 dB back-off output power level with 3.43% improved drain efficiency, while the peak power is reduced by 0.88 dB. This reduced power is the consequence of the reduced drain bias of the carrier amplifier. Fig. 3.5C presents the gain profiles versus the output power level for each amplifier and overall Doherty amplifier. Because the GaN HEMT device has a low gm at the low $V_{dc}$ region, the carrier amplifier of the modified Doherty amplifier has relatively low gain compared with the conventional carrier amplifier. On the other hand, the peaking amplifier delivers the same fundamental output power and gain, comparable with the conventional Doherty case. Accordingly, the overall gain flatness versus the output power level of the modified Doherty amplifier is better than the other case. In spite of the lower gain and output power level, the modified Doherty amplifier delivers improved efficiency as shown in Fig. 3.5B.

Fig. 3.6 shows the simulated load impedances and the fundamental drain currents of the carrier and peaking amplifiers versus the output power level, respectively. When the peaking amplifiers of both the Doherty amplifiers are turned on at the 6 dB back-off output power level as shown in Fig. 3.6B, the modified Doherty amplifier converges properly to the 50 Ω load impedance, delivering the proper load modulation, while the conventional one does not.

In summary, the modified Doherty amplifier employing asymmetrical $V_{dc}$ is a very useful technique for the proper load modulation with first peak efficiency at the 6 dB back-off point and the flatter gain versus the output power level. However, to take the advantages, the peak power is reduced because of the lower drain bias of the carrier amplifier. Therefore, the Doherty amplifier using the optimized offset lines, which has a larger load modulation ratio, may be a preferred technique than the modified Doherty amplifier described in this section. This asymmetrical drain bias technique can be employed to optimize the first peak efficiency point of a Doherty amplifier. The possibility is shown in Fig. 3.3.

## 3.2 OPTIMIZED DESIGN OF GaN HEMT DOHERTY POWER AMPLIFIER WITH HIGH GAIN AND HIGH EFFICIENCY

Theoretically, the gain of a carrier amplifier having $2R_{OPT}$ should be 3 dB higher than that with $R_{OPT}$. However, the nonlinear input capacitance of the device affects the input match, and the gain can be different according to the operation point. For a proper load modulation with a high gain and high power-added efficiency (PAE), as we describe

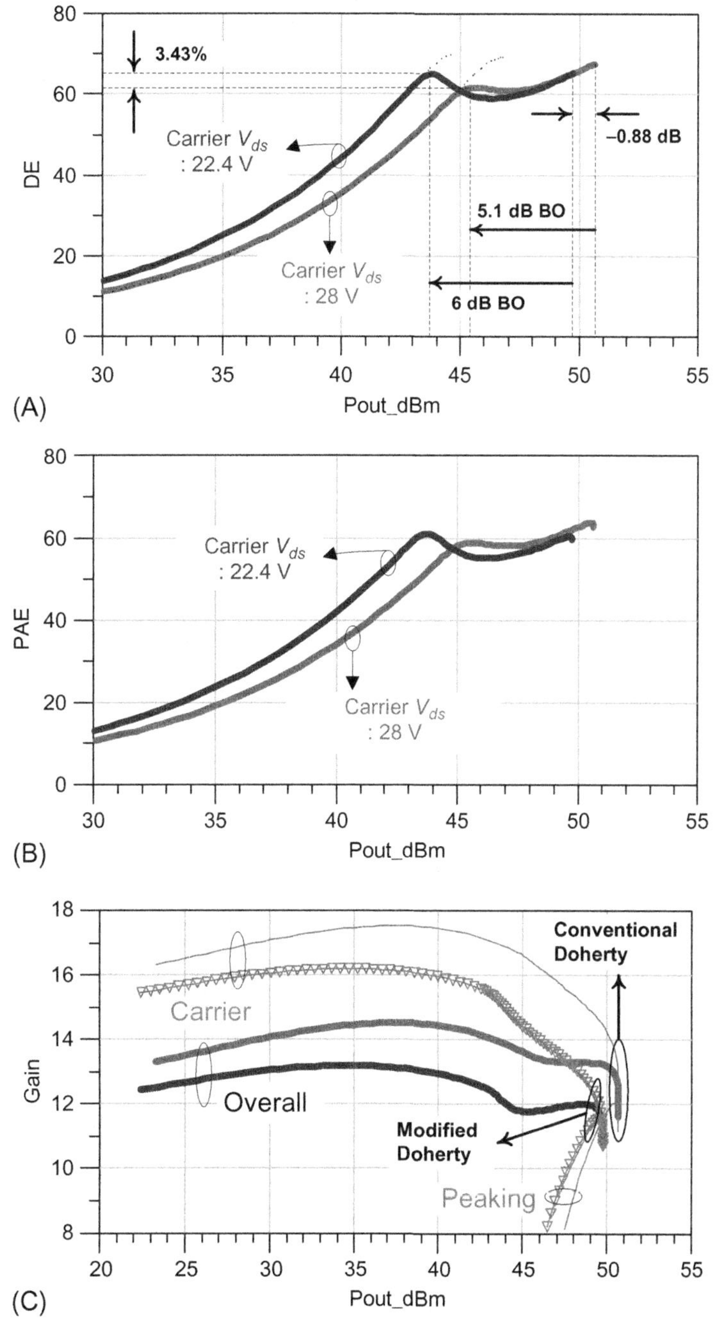


**Fig. 3.5** ADS simulation results of the conventional Doherty and the modified Doherty amplifier: (A) drain efficiency, (B) power-added efficiency, and (C) gain characteristics versus output power level.

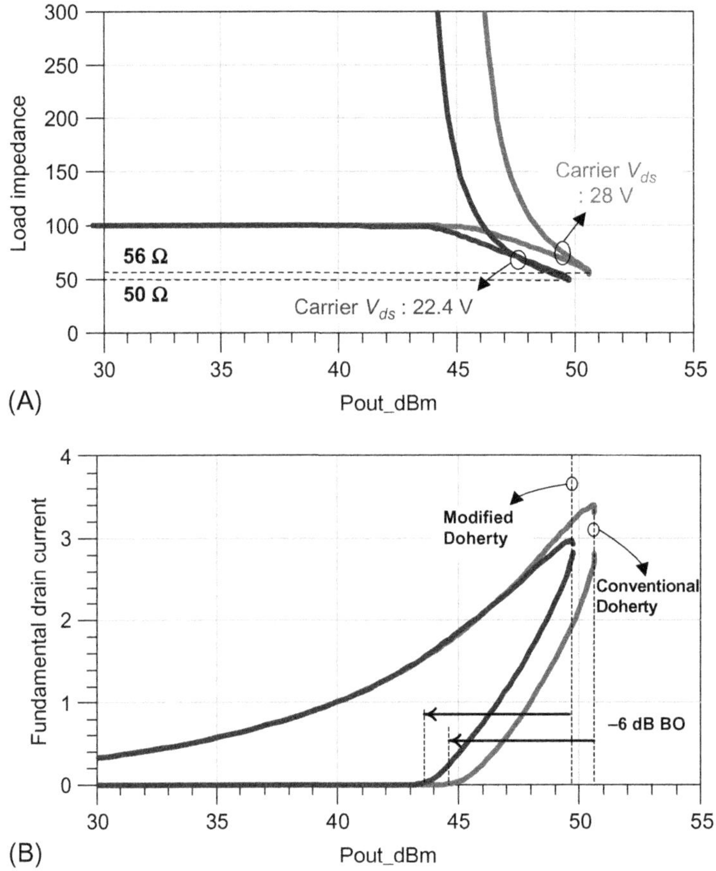


**Fig. 3.6** ADS simulation results of the conventional Doherty and the modified Doherty amplifier: (A) load impedances and (B) DC current versus output power level of the carrier and peaking amplifiers.

in the previous section, the input of the carrier amplifier should be matched at the first peak power region with $2R_{\text{OPT}}$ and that of peaking at the peak power region. These optimized design based on the GaN HEMT device is introduced.

### 3.2.1 Optimized Design of Carrier and Peaking Amplifiers

In a conventional Doherty architecture, the input of the carrier amplifier is conjugate-matched at the peak power with the output load of $R_{\text{OPT}}$. This match produces mismatch at a back-off power level and degrades the gain and the PAE. To solve the problem, the input should be matched at the first peak efficiency point.

Fig. 3.7A shows the nonlinear equivalent circuit model of the GaN HEMT. Here, we assume that the current source sees the load $R_L$ for a power match. The extracted input capacitance, $C_{\text{in}}$, of the model is given by

$$C_{\text{in}} = C_{gs} + C_{gd}(1 + g_m R_L) \tag{3.11}$$

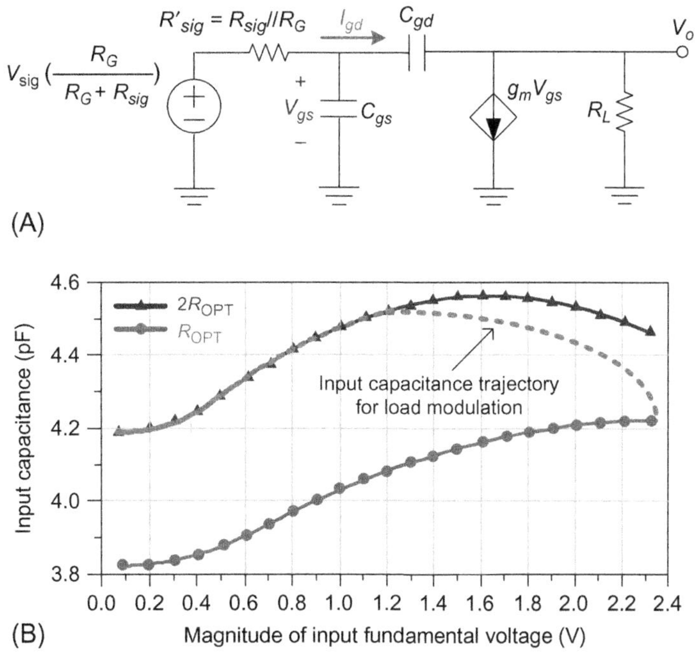


**Fig. 3.7** (A) GaN HEMT nonlinear equivalent circuit model. (B) Extracted input capacitance $C_{in}$ for $R_{OPT}$ and $2R_{OPT}$ loads.

The $C_{gs}$ increases as the input power increases. The output load, $R_L$, of the carrier amplifier decreases from $2R_{OPT}$ to $R_{OPT}$ through the load modulation. As a result, the input capacitance, $C_{in}$, of the carrier amplifier is varied as the input power increases as depicted in Fig. 3.7B. This $C_{in}$ variation affects the power matching and gain.

Fig. 3.8 shows the source-pull simulation results for the fundamental impedance matching points of the carrier amplifier for the output load impedances of $R_{OPT}$ and $2R_{OPT}$. The input power level for $R_{OPT}$ is 6 dB higher than that for $2R_{OPT}$. As the load impedance is modulated from $2R_{OPT}$ to $R_{OPT}$ with the increased input power, the effective input capacitance decreases. By matching the input impedance at the first peak power level with the $2R_{OPT}$ load, the gain of the carrier amplifier at the back-off region can be matched accurately with high efficiency although the matching tolerance is small. The slight mismatch at the lower power region (below the first peak power) provides a flat gain of the Doherty amplifier because the gain at the lower power region is higher than that at the first peak power due to the smaller $C_{in}$. The power match at the higher power operation is maintained because the optimum impedance contour at the high power region is insensitive due to the saturated operation with large tolerance as shown in the figure.

In Fig. 3.9, the CW characteristics of the first peak input matched carrier amplifier is compared with the peak power matched conventional amplifier. The new design allows a

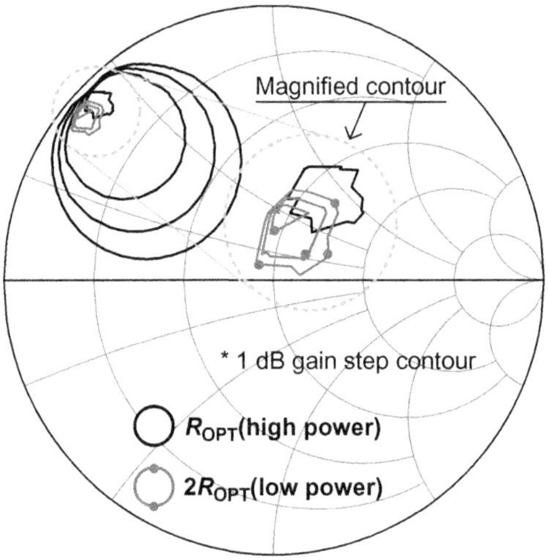


**Fig. 3.8** Source-pull simulation results for fundamental matching impedances of the carrier amplifiers for $R_{OPT}$ and $2R_{OPT}$ loads.

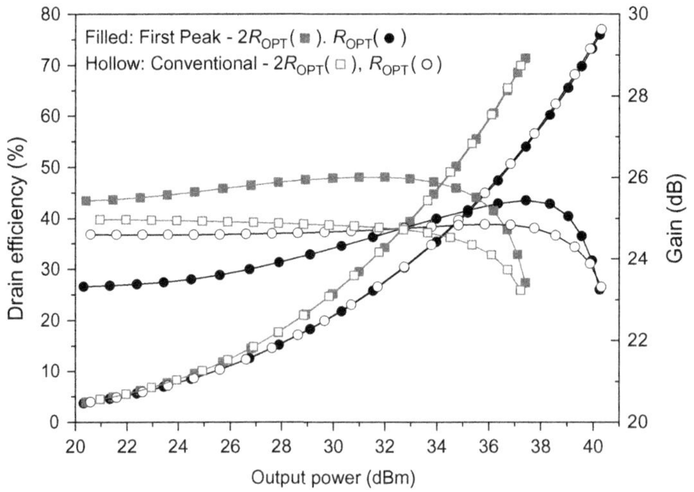


**Fig. 3.9** Simulated CW characteristics of the first peak matched and conventional carrier amplifiers for the $R_{OPT}$ and $2R_{OPT}$ loads.

proper Doherty modulation since the gain difference between the $2R_{OPT}$ and $R_{OPT}$ matches is about 3 dB. However, the conventional design can achieve the load modulation through the saturated operation of the peaking amplifier only because the two gains are almost identical. This load modulation through the saturated operation is described in Section 5.2.

The input of the peaking amplifier is conjugate-matched at the high power level to maximize the gain at the peak power level. The harmonics of the output impedance for

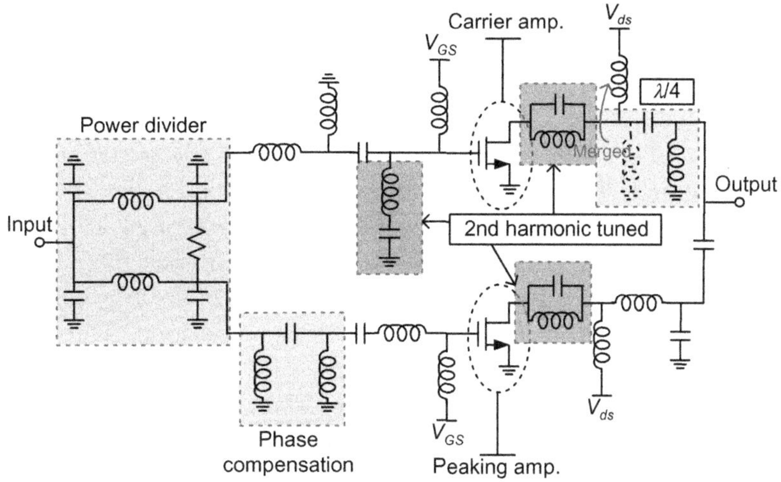


**Fig. 3.10** Schematic of the first peak input matched Doherty amplifier.

the carrier and peaking amplifiers are matched to the optimum impedances for a high saturated drain efficiency, using LC parallel network circuits. The design method of the saturated amplifier can be found at the further reading. Fig. 3.10 represents the schematic of the proposed Doherty amplifier, which is implemented in MMIC.

## 3.2.2 Operation of the Optimally Matched Doherty Amplifier

For the simple circuit topology in MMIC, the fundamental output impedance at the output node of the carrier amplifier increases from 50 to 100 Ω, instead of the conventional design of 25 to 50 Ω, and that of the peaking amplifier decreases from infinity to 100 Ω as shown in Fig. 3.11A. For the optimum Doherty operation, the input power driving

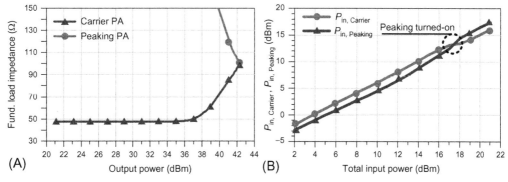

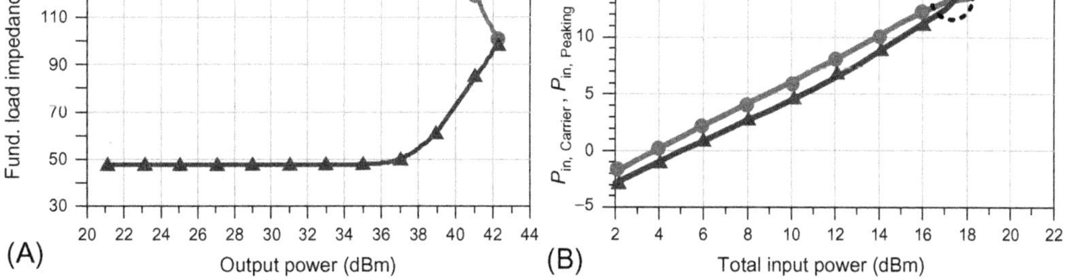


**Fig. 3.11** (A) Fundamental output load impedances for the carrier and peaking amplifiers. (B) Uneven input power dividing achieved by the first peak input matching.

needs to be an uneven drive. The uneven input dividing is normally realized using the uneven power divider. It can be achieved using the nonlinear input capacitance of the peaking amplifier also as described in Section 5.3.1. Similarly to the nonlinear capacitance case, we could easily realize the uneven input dividing using the input matching. As shown in Fig. 3.11B, 1 dB more power is delivered to the carrier amplifier at the low-power operation and to the peaking amplifier at the high power. Fig. 3.12 shows a CW simulation result of the Doherty amplifier implemented at 1.8 GHz using TriQuint 3MI 0.25 m gallium nitride (GaN) high-electron-mobility transistors (HEMT) MMIC. The DE is more than 52% at the 6 dB back-off power and 59% at the peak power without degradation during the load modulation. Through the optimized input matching for the carrier amplifier, the gain at the back-off power is high for the single stage amplifier, over 19.5 dB. Because of the high gain of the carrier amplifier at the back-off power level, the gate bias voltage for the peaking amplifier can be increased, preventing the gain degradation of the Doherty amplifier at the modulation region as shown in Fig. 3.12.

The implemented integrated circuit size except for the RF choke inductor is 2.28 × 3.09 mm as shown in Fig. 3.13. Drain bias voltage is 28 V for both the carrier and peaking amplifiers, and gate bias voltages are −2.8 V for the carrier amplifier and −3.26 V for the peaking amplifier. The Doherty amplifier is tested using an LTE signal with 10 MHz bandwidth and 6.5 dB PAPR at 1.8 GHz. The Doherty amplifier achieves a drain efficiency of 57.2% (PAE of 56.3%) and a gain of 18.8 dB at an average output power of 35.6 dBm.

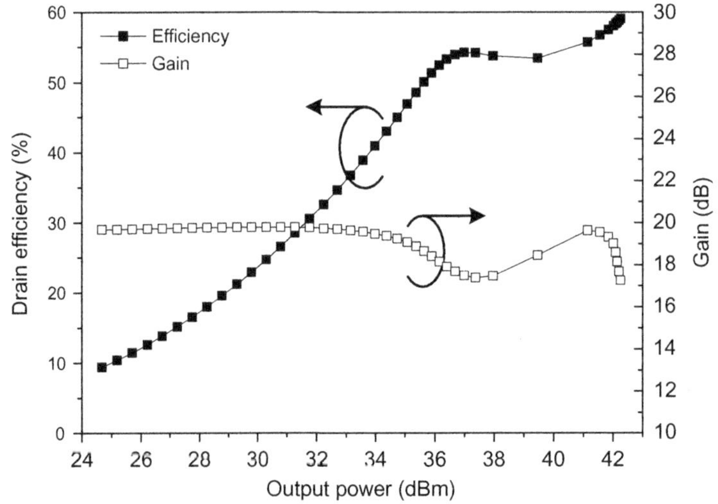


Fig. 3.12 Simulated CW characteristics of the Doherty amplifier.

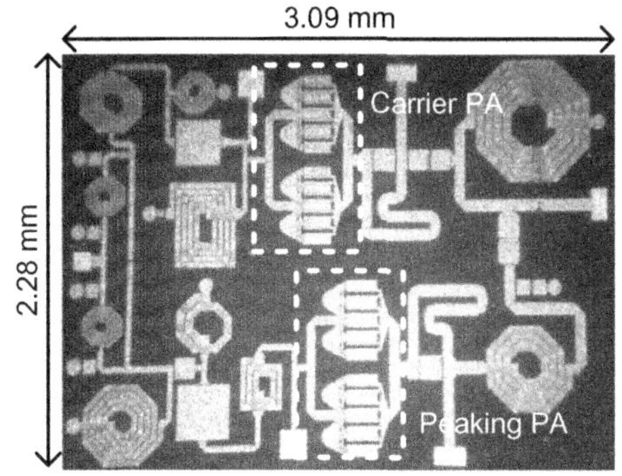


**Fig. 3.13** Photograph of the fabricated amplifier chip.

## 3.3 OPTIMIZED PEAKING AMPLIFIER DESIGN FOR DOHERTY AMPLIFIER

The peaking amplifier operates at a class C bias and has a lower maximum output power than that of the carrier amplifier. Because of the low output power of the peaking amplifier, the load modulation of the Doherty network cannot be properly carried out. To solve the problem, an inductive second harmonic load is employed at the input to generate the out-phased second harmonic component. The out-phased second harmonic flattens the sign wave, and the conduction angle of the input voltage waveform is enlarged, restoring the decreased conduction angle and the output power. In addition, the peaking amplifier can be turned on at a higher input power due to the reduced peak current, and the first peak efficiency of the Doherty amplifier is enhanced.

### 3.3.1 Optimized Design of Peaking Amplifier for Proper Doherty Operation

For the proper load modulation of the Doherty amplifier, the output currents of the carrier and peaking amplifiers should be increased with the input voltage as shown in Fig. 3.14. To get the current prolife, as we have discussed earlier, the transconductance of the peaking amplifier should be two times larger than that of the carrier amplifier. To minimize the production cost, however, the device sizes of the carrier and peaking amplifiers are the same since they generate the same maximum power. To overcome the lower power problem, the uneven input drive or the gate bias adaptation technique can be employed.

The problem can be solved by the input second harmonic voltage control, also. Usually, the input second harmonic load of a power amplifier is shorted for a small conduction angle of the output current, which is required for a high-efficiency operation. However, we can utilize the input second harmonic voltage generated from the nonlinear input capacitor to increase the conduction angle and thereby increase the

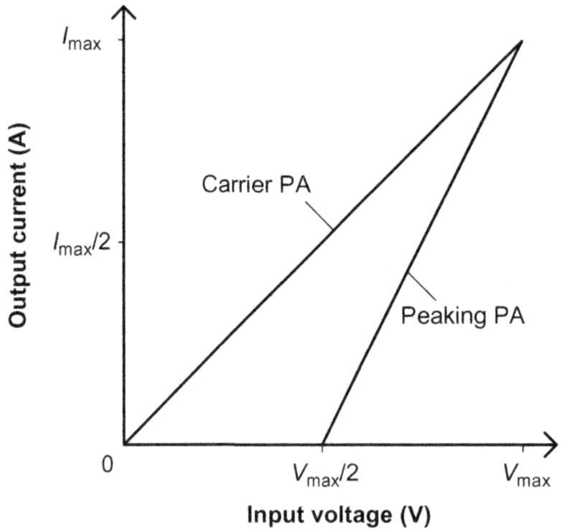


**Fig. 3.14** Output currents versus input voltage of the carrier and peaking amplifiers for an ideal Doherty operation.

output power. The input capacitance variation for the bias voltage has a reverse characteristic to the output capacitor, that is, it increases as the gate-source voltage increases as shown in Fig. 3.15A, and it generates the out-phased second harmonic voltage. In contrast to the in-phased second harmonic voltage, the out-phased second harmonic voltage flattens the peak of the voltage waveform as shown in Fig. 3.15B. With the flattened input voltage waveform, the effective conduction angle increases, compensating the reduced conduction angle of the peaking amplifier due to the class C bias. In addition, the peak value of the input voltage decreases, allowing the peaking amplifier to turn on late at the same gate bias condition. Therefore, the peaking amplifier with the peak flattened input voltage waveform is suitable for the ideal Doherty operation. To take full advantage of the nonlinear input capacitor, the input second harmonic load should be an inductive load. As shown in Fig. 3.15C, the peak input voltage decreases with the properly tuned inductive second harmonic load.

### 3.3.2 Simulation and Experimental Results

To validate the inductive harmonic tuning concept, a Doherty amplifier is designed at 2.14 GHz using TriQuint 0.25 μm GaN HEMT process. With the inductive input second harmonic load for the peaking amplifier, the input voltage has a peak flattened waveform. Due to the voltage waveform, the conduction angle of the peaking amplifier increases, and the peak input voltage is decreased as shown in Fig. 3.16.

Fig. 3.17 shows the CW performances of the designed carrier and peaking amplifiers. The input second harmonic load of the carrier amplifier is shorted, and the peaking amplifiers are designed with the short and inductive loads. $V_{GS}$ is −3 V for the carrier amplifier and −4.5 V for the peaking amplifier. With the second harmonic short, the

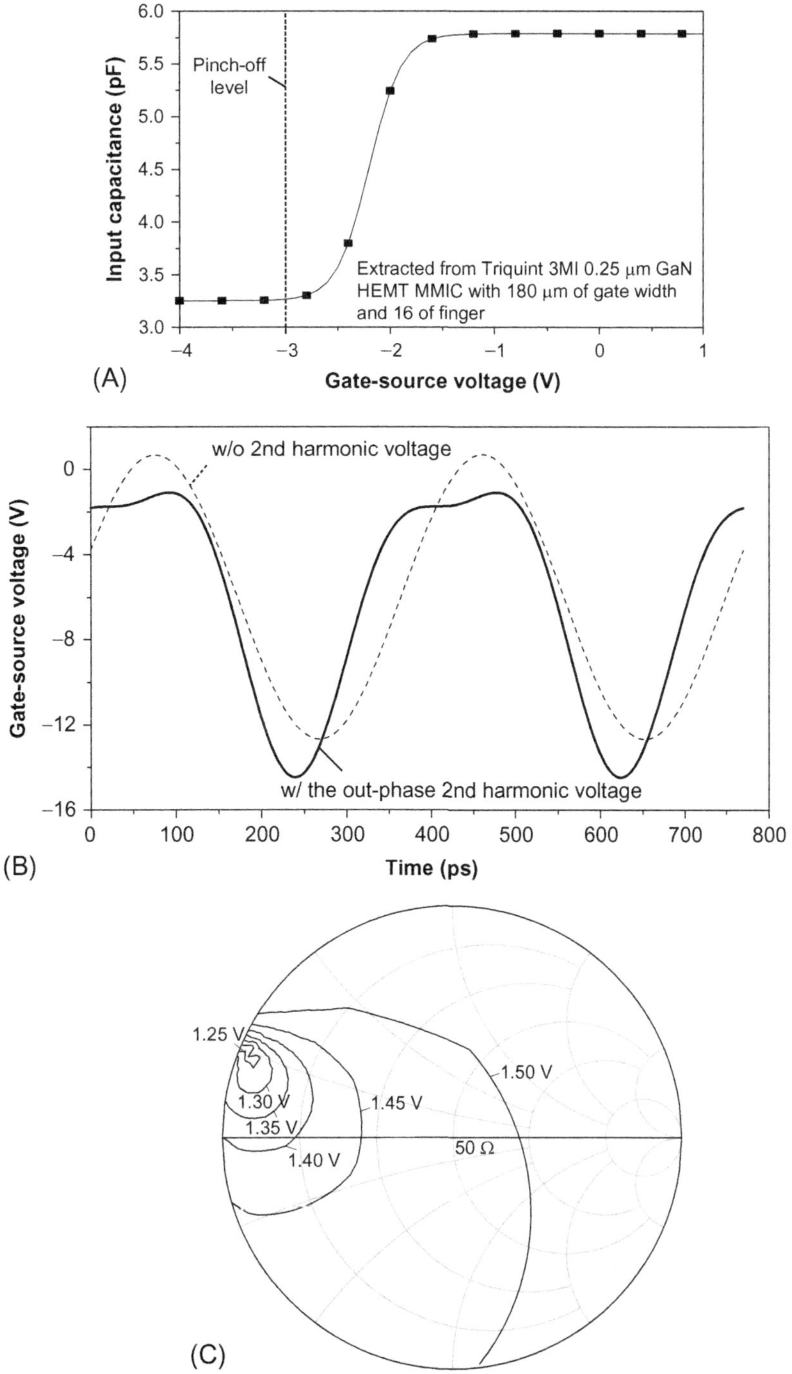


**Fig. 3.15** (A) Nonlinear input capacitance according to the gate-source bias voltage. (B) Input voltage waveform with and without the out-phased second harmonic voltage. (C) Second harmonic source-pull simulation results for the peak input voltages.

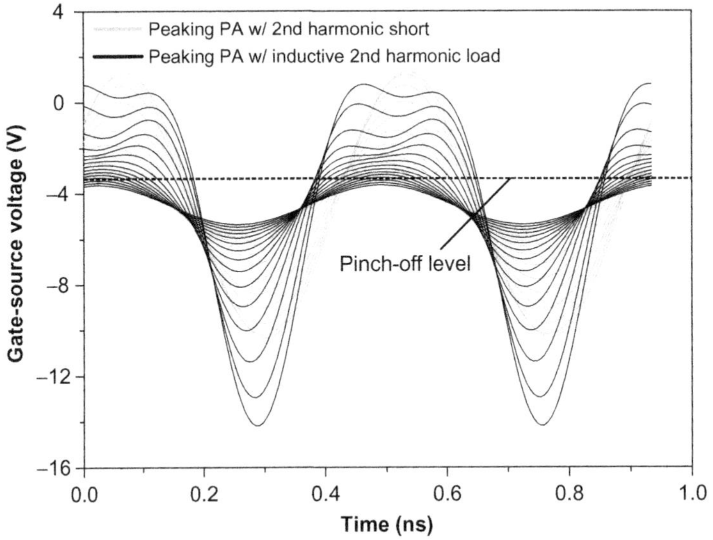


**Fig. 3.16** Input voltage waveforms of the peaking amplifier with the short and inductive input second harmonic load.

peaking amplifier delivers a lower output power and a higher efficiency than the carrier amplifier. However, with the inductive input second harmonic load, the peaking amplifier generates a larger output power, comparable with the carrier amplifier. In addition, with the same gate bias condition, the peaking amplifier with the inductive input second harmonic load turns on later than that of the peaking amplifier with the input second harmonic short circuit. These CW characteristics of the peaking amplifier are beneficial for the Doherty amplifier operation. Although the efficiency is degraded a little, the power is more important for the peaking amplifier.

Fig. 3.18 shows the schematic of the Doherty amplifier adopting the second harmonic tuned peaking amplifier. To realize the inductive input second harmonic load, an inductor is added after the second harmonic short circuit at the input of the peaking amplifier. The device size of the peaking amplifier is the same as that of the carrier amplifier. $V_{GS}$ of the peaking amplifier is increased to $-4.3$ V to get the same turn-on voltage. Fig. 3.19 shows the two-tone performance of the Doherty amplifier. At the output power of 37 dBm, the IMD3s at the two sides are lower than $-35$ dBc, and the power-added efficiency is 47% (including the DC power consumption of the drive stage).

Fig. 3.20 shows the fabricated Doherty amplifier based on the schematic in Fig. 3.18. The integrated circuit size except for the RF choke inductor is $2.3 \times 3.5$ mm. The drain bias voltages are 28 V for the power stage of the carrier and peaking amplifiers and 20 V for the drive amplifier. The gate bias voltages of the drive amplifier and the carrier amplifier are $-2.8$ V and that for the peaking amplifier is $-4.2$ V. Fig. 3.21 shows the measured two-tone performance of the fabricated Doherty amplifier, which is similar to the

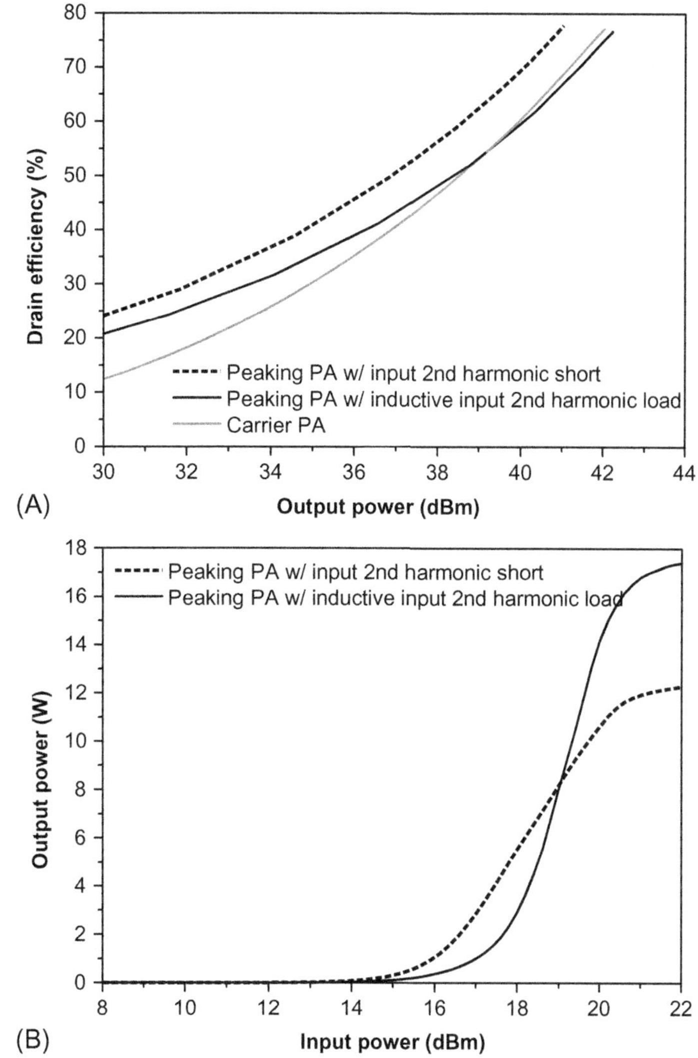


**Fig. 3.17** (A) Simulated CW performances of the carrier and peaking amplifiers. (B) Output power versus input power of the peaking amplifiers.

simulation one. Fig. 3.22 shows the performance of the fabricated Doherty amplifier for the LTE signal at 2.14 GHz with a 10 MHz bandwidth and 6.5 dB PAPR. At the output power of 46.7 dBm, ACLR is −35 dBc, PAE is 46.7% (including the DC power consumption of the drive stage), and gain is 29.3 dB.

In conclusion, the peaking amplifier cannot deliver the same maximum output power as that of the carrier amplifier because of the low gate bias voltage of the peaking amplifier and cannot properly modulate the load like an ideal Doherty amplifier. To solve the

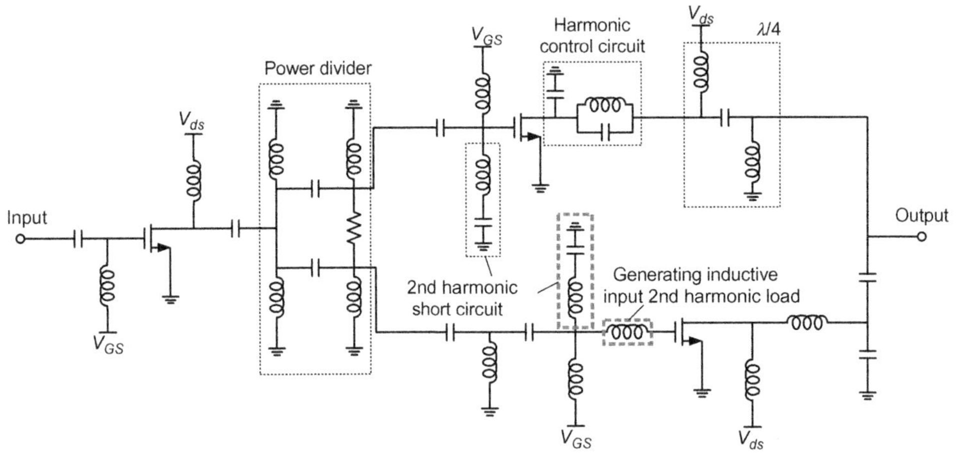


**Fig. 3.18** Schematic of the second harmonic tuned Doherty amplifier.

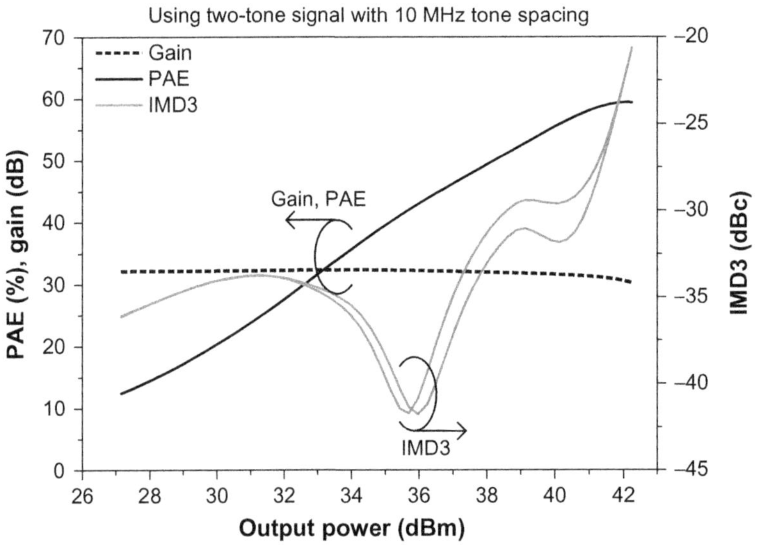


**Fig. 3.19** Simulated two-tone performance of the Doherty amplifier.

problem, the inductive input second harmonic load is employed. With the harmonic load, the input voltage becomes the peak flattened waveform, and it enlarges the conduction angle of the peaking amplifier. Therefore, the reduced conduction angle of the peaking amplifier, due to the low gate bias, is restored, and the proper Doherty operation could be realized.

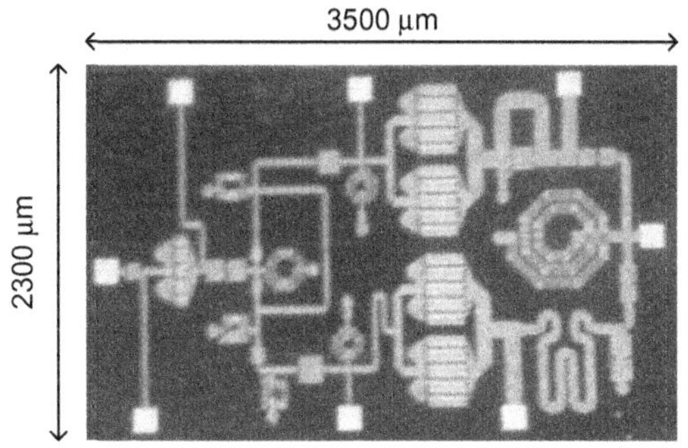


**Fig. 3.20** Photo of the implemented Doherty amplifier.

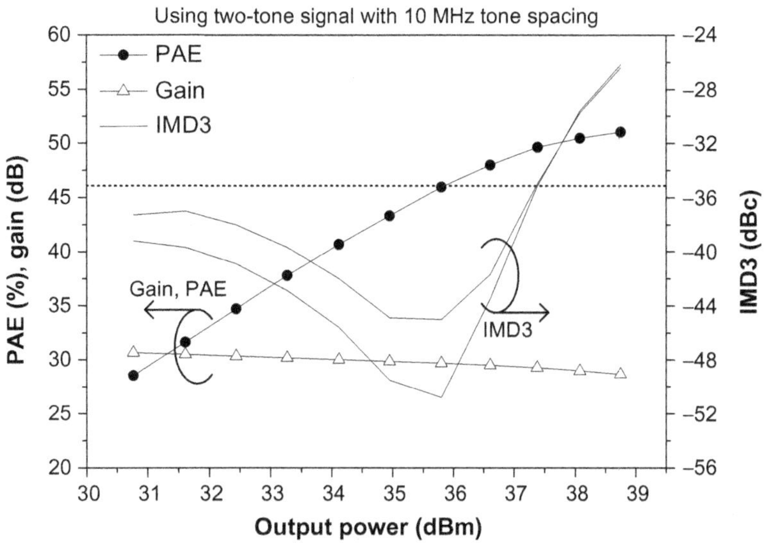


**Fig. 3.21** Measured two-tone performance of the fabricated Doherty amplifier.

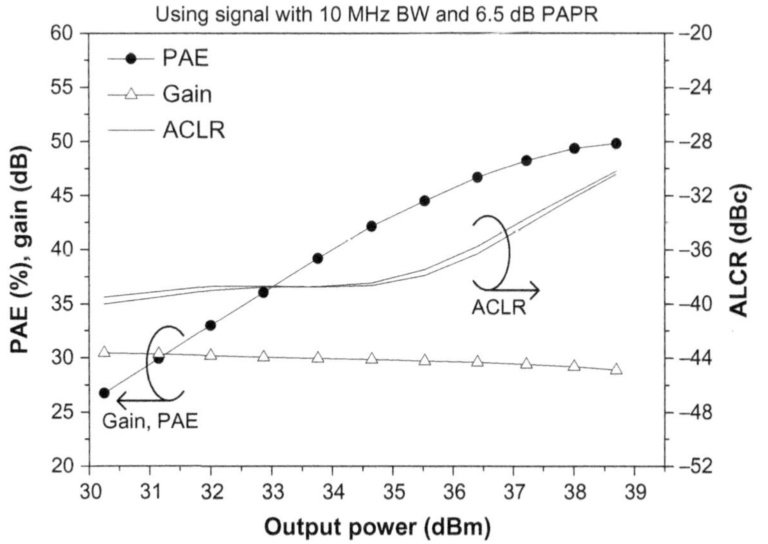


**Fig. 3.22** Measured LTE performance of the fabricated Doherty amplifier.

## 3.4 SATURATED DOHERTY AMPLIFIER

A Doherty amplifier is based on the carrier amplifier with a deep class AB bias and the peaking amplifier with a class C; thus, the maximum efficiency of 78.5% is expected. However, the efficiency can be even higher by employing a switching amplifier as an unit amplifier but at the expense of linearity. A class F or $F^{-1}$ amplifier is a good choice since the harmonic-controlled matching topology could provide higher efficiency. To get the desired efficiency, the fundamental, second, and third harmonic impedances should be properly controlled simultaneously. In this section, a saturated Doherty amplifier design using the class F amplifier topology is introduced to maximize efficiency. But class $F^{-1}$, class E, or other switching/saturated amplifiers can be designed similarly.

The unit amplifier should be operated at the saturated mode, and the Doherty amplifier is named as a saturated Doherty amplifier. The fully matched output matching network is discussed in this section, including the harmonic control circuit (HCC) for the saturated operation. The operational behavior of the saturated Doherty amplifier is analyzed with finite harmonic contents, including the harmonic-controlled load modulation behavior, efficiency, and linearity.

### 3.4.1 Operational Principle of the Saturated Doherty Amplifier

The efficiency of the Doherty amplifier can be enhanced by employing the carrier and peaking amplifiers based on the class $F^{-1}$ mode. Fig. 3.23 shows the schematic diagram of the fully matched saturated Doherty amplifier including the HCC and offset lines.

Fig. 3.23A shows an operational diagram of the saturated Doherty amplifier. For the Doherty operation, the carrier amplifier is biased at the pinch off and the peaking amplifier below the pinch off. The current and voltage waveforms at the devices of the carrier and peaking amplifiers are shaped by the harmonic impedances of the HCC. Theoretically, the high efficiency can be achieved in class F operation by generating half-sinusoidal current and square-wave voltage waveforms. These waveforms can be realized by creating the zero impedances at all even harmonics and infinite impedances at all odd harmonics. However, in a practical design, all harmonic contents cannot be controlled. Moreover, the amplifiers do not operate in the saturated mode for all power levels, either. Thus, a HCC, which can control the second and third harmonics, has been employed in front of the output matching circuit since the HCC does not need to be modulated. The Doherty circuit topology is depicted in Fig. 3.24. The HCC, which is shown in Fig. 3.25, shows the open and short circuits for the class F harmonic tuning. The harmonic matching impedances should not be changed with the matching and load modulation circuits. For the purpose, the second and third shorts are introduced after cutting off the impedance modulation effect. For proper Doherty operation, the offset line at the back of the output matching circuit has also been employed. Therefore, the harmonic control circuit is attached right at the

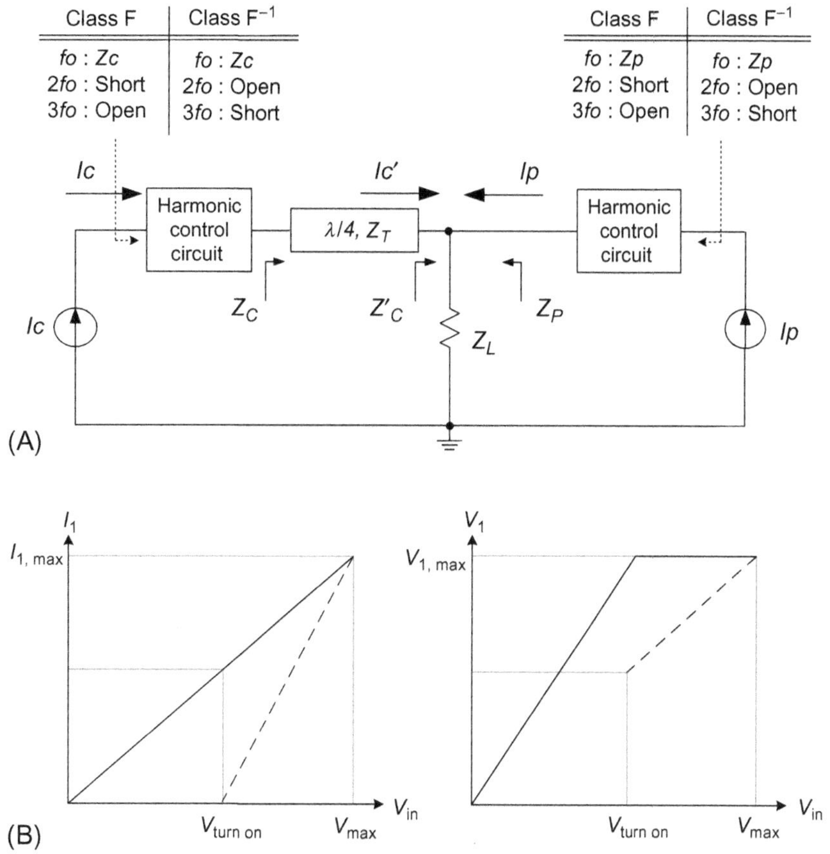


**Fig. 3.23** (A) Operational diagram of the saturated Doherty amplifier. (B) Fundamental current and voltage versus input drive voltage (*solid line*, carrier amplifier, and *dot line*, peaking amplifier).

drain terminal, and a tuning line is added for compensating the harmonic detuning effect by the devices' output capacitance. The amplifier circuit topology is depicted in Fig. 3.25.

The analysis of the circuit is based on the following assumptions:

**(1)** Each current source is linearly proportional to the input voltage above the pinch-off voltage.

**(2)** The voltage waveform depends only on the fundamental and third harmonic components.

**(3)** The maximally flat voltage waveform is generated by a 1/9 amplitude ratio and 180 phase difference between the fundamental and third harmonic component voltages, which is the voltage waveform of a class F amplifier.

**(4)** The fundamental load impedance is properly modulated using the uneven power drive.

Fig. 3.23B represents the magnitudes of the fundamental current and voltage according to the input drive voltage. Since the current and voltage waveforms represent the

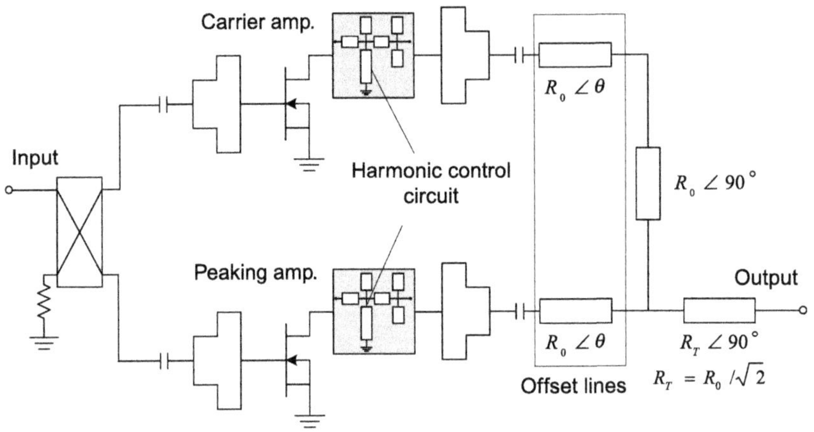


**Fig. 3.24** Schematic diagram of the saturated Doherty amplifier.

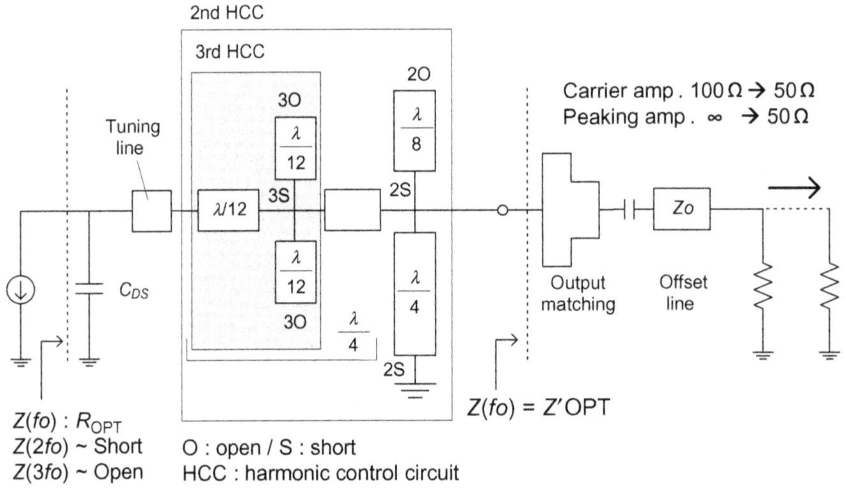


**Fig. 3.25** Fully matched output matching network including the HCC.

half-sinusoidal current and maximally flat voltage waveforms, $I_{1,\max}$ and $V_{1,\max}$ are a half of $i_{\text{peak}}$ and 9/8 of $V_{DC}$, respectively. Here, $I_{1,\max}$ and $V_{1,\max}$ are the maximum fundamental components of the current and voltage, respectively; $i_{\text{peak}}$ is the maximum current, and $V_{DC}$ is the dc bias voltage. At the low-power region ($0 < V_{\text{in}} < V_{\text{turn on}}$), where only the carrier amplifier is active, the fundamental and dc currents of the amplifier are given by

$$I_C = I_{1,\max}\left(\frac{v_{\text{in}}}{V_{\max}}\right) \tag{3.12}$$

$$I_{DC,C} = I_{dc,c}\left(\frac{v_{\text{in}}}{V_{\max}}\right) \tag{3.13}$$

where $I_{dc,c}$ is $1/\pi$ of $i_{\text{peak}}$ and $V_{\max}$ is the maximum input voltage.

To analyze the load modulation behavior of the saturated Doherty amplifier, we use $Z_T = R_{OPT}$ and $Z_L = R_{OPT}/2$, in Fig. 3.23A. The load impedance of the carrier amplifier $Z_C$ is given by

$$Z_C = \frac{Z_T^2}{Z_L} = 2R_{OPT} \quad \text{where,} \quad R_{OPT} = \frac{9}{8}\left(\frac{V_{DC}}{I_{1,\max}}\right) \tag{3.14}$$

Due to the maximally flat voltage waveform, the fundamental voltage component is 9/8 times larger impedance than that of the class B amplifier. Thus, the load impedance of each amplifier is designed to have 9/8 times larger impedance than the conventional amplifier.

At the higher power region, both the carrier and peaking amplifiers are active. The amplifier is unevenly driven with the power ratio $\sigma$, which is defined in Eq. (2.4):

$$\sigma = \frac{I_{1,\max}}{I_{1,p}(1-\psi)} \tag{3.15}$$

where

$$\psi = \frac{V_{\text{turn on}}}{V_{\max}} \tag{3.16}$$

$$I_{1,p} = \frac{I_{\max}}{2\pi} \cdot \frac{\theta_p - \sin\theta_p}{1 - \cos(\theta_p/2)} \tag{3.17}$$

Here, $\psi$ is the turn-on portion, and $I_{1,\ p}$ is the fundamental component of the peaking current with a conduction angle of $\theta_p$.

The fundamental and dc currents of the peaking amplifier are expressed as

$$I_P = \sigma I_{1,p}\left(\frac{v_{\text{in}}}{V_{\max}} - \psi\right) \tag{3.18}$$

$$I_{DC,P} = \sigma I_{dc,p}\left(\frac{v_{\text{in}}}{V_{\max}} - \psi\right) \tag{3.19}$$

$$I_{dc,p} = \frac{I_{\max}}{2\pi} \cdot \frac{2\sin(\theta_p/2) - \theta_p\cos(\theta_p/2)}{1 - \cos(\theta_p/2)} \tag{3.20}$$

The load impedances of the carrier and peaking amplifiers are given by

$$Z_C = \frac{Z_T^2}{[Z_L \cdot (1 + I_P/I'_C)]} \tag{3.21}$$

$$Z_P = Z_L\left(1 + \frac{I'_C}{I_P}\right) \tag{3.22}$$

The fundamental load impedance of the carrier amplifier is modulated from $2R_{OPT}$ to $R_{OPT}$, and the load impedance of the peaking amplifier modulated from $\infty$ to $R_{OPT}$ while the second harmonic short and third harmonic open impedances are maintained.

### 3.4.2 Efficiency and Linearity of the Saturated Doherty Amplifier

The power-level-dependent operation is explored by defining the saturation state and the unsaturation state according to the amplitude ratio/phase difference between the fundamental and third harmonic voltage components. The amplitude ratio $\alpha$ and the relative phase difference $\beta$ are expressed as

$$\begin{aligned} V(t) &= V_{DC} + V_1 \cos(\omega t + \phi_1) + V_3 \cos(3\omega t + \phi_3) \\ &= V_{DC} + V_1 \cos(\phi_1) + V_3 \cos(\phi_3) \end{aligned} \tag{3.23}$$

$$\alpha = \left| \frac{V_3}{V_1} \right| \tag{3.24}$$

$$\beta = \phi_1 - \phi_3 \tag{3.25}$$

where $V_1$ and $V_3$ are magnitudes of the fundamental and third harmonic voltages, respectively. $\phi_1$ and $\phi_3$ are phases of the fundamental and third harmonic voltages, respectively. At the saturation state, the amplifier should generate the half-sinusoidal current and maximally flat voltage waveforms with $\alpha = 1/9$ and $\beta = 180°$ at the given drive level. $\alpha$ and $\beta$ values of the saturated Doherty amplifier have been investigated using the ADS simulator. Based on the data, we have analyzed the efficiency and the linearity.

#### *3.4.2.1 Efficiency of the Saturated Doherty Amplifier*

The efficiency analysis can be carried out using only the fundamental and dc components according to the current and voltage waveforms. At the low-power region $(0 < V_{in} < V_{turn\ on})$ where only the carrier amplifier is turned on, the RF and dc powers are given by

$$\begin{aligned} P_{RF} &= \frac{1}{2} I_C^2 \cdot Z_C \\ &= \frac{I_{1,c}^2}{2} \left( \frac{v_{in}}{V_{max}} \right)^2 \cdot 2Z_0 \\ &= I_{1,c} \left( \frac{v_{in}}{V_{max}} \right)^2 \cdot \left( \frac{9}{8} \right) V_{DC} \end{aligned} \tag{3.26}$$

$$\begin{aligned} P_{DC} &= I_{DC,C} \cdot V_{DC} \\ &= I_{dc,c} \left( \frac{v_{in}}{V_{max}} \right) \cdot V_{DC} \end{aligned} \tag{3.27}$$

From Eqs. (3.26), (3.27), the efficiency becomes

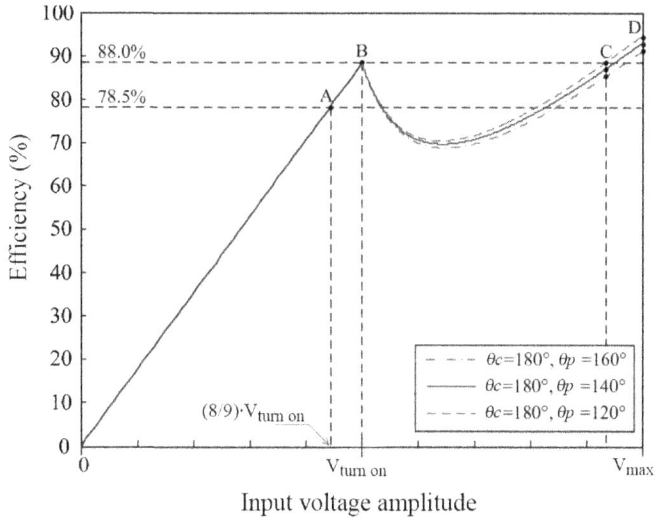


**Fig. 3.26** Efficiency versus input drive level according to the bias point of the peaking amplifier.

$$\eta = \frac{P_{RF}}{P_{DC}} \times 100 = \frac{I_{1,c}}{2I_{dc,c}} \cdot \frac{9}{8}\left(\frac{v_{\text{in}}}{V_{\text{turn on}}}\right) \times 100 \tag{3.28}$$

The carrier amplifier operates at the unsaturated state until the (8/9) $V_{\text{turn on}}$ input drive level. At this level, the amplifier has a class B peak efficiency due to the larger load impedance as shown in Fig. 3.26. The efficiency increases further to that of the third harmonic tuned class F amplifier as the amplifier is pushed into saturation. The peaking amplifier operates similarly. The efficiency of the peaking amplifier is slightly higher with deeper class C operation.

The load lines are slightly elliptical at a low drive level, and the load line of the carrier amplifier reaches to the minimum allowable voltage at the peaking turn-on point, as shown in Fig. 3.27A where the turn-on voltage of the peaking amplifier ($V_{\text{turn on}}$) is

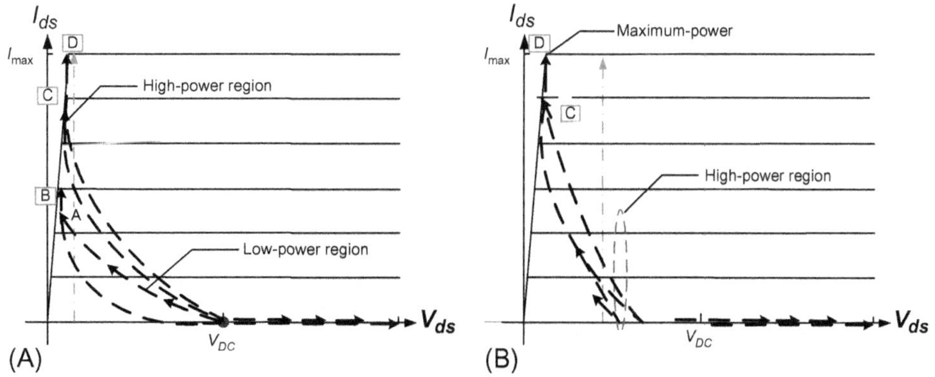


**Fig. 3.27** Load lines of two amplifiers: (A) carrier amplifier and (B) peaking amplifier.

set to $V_{max}/2$. The amplifier reaches to the saturation state above this drive level, and as shown in Fig. 3.26, the efficiency increases from the class B peak efficiency (point A) to the class F peak efficiency (point B). The load line changes from the slightly elliptical to the quasi-"$L$" curve at the turn-on voltage, as shown in Fig. 3.27A, but the line cannot make the perfect "$L$" curve because the harmonic voltage control is limited to the third harmonic. This state illustrates that the amplifier has the half-sinusoidal current and maximally flat voltage waveforms, and the efficiency can reach to the peak efficiency of the class F amplifier with third harmonic tuning, i.e., 88%.

At the higher power region ($V_{turn\ on} < V_{in} < V_{max}$), both amplifiers are active. The RF and dc powers are sum of the two amplifiers' and are given by

$$P_{RF} = \frac{1}{2}\left(I_C^2 Z_C + I_P^2 Z_P\right)$$
$$= \frac{V_{DC}}{2}\left(\frac{9}{8}\right)\left[\frac{2I_{1,max}^2 V^3}{\left(I_{1,max} + \sigma I_{1,p}\right)V - \sigma I_{1,p}\psi} + \frac{\sigma I_{1,p}(V-\psi)\left\{\left(I_{1,max} + \sigma I_{1,p}\right)V - \sigma I_{1,p}\psi\right\}}{2I_{1,max}}\right] \tag{3.29}$$

$$\begin{aligned} P_{DC} &= \left(I_{DC,C} + I_{DC,P}\right) \cdot V_{DC} \\ &= \left[I_{dc,c} V + \sigma I_{dc,p}(V-\psi)\right] \cdot V_{DC} \end{aligned} \tag{3.30}$$

From Eqs. (3.29), (3.30), the efficiency becomes

$$\eta = \frac{P_{RF}}{P_{DC}} \times 100$$
$$= \frac{1}{2}\left(\frac{9}{8}\right)\frac{\left[\frac{2I_{1,max}^2 V^3}{\left(I_{1,max} + \sigma I_{1,p}\right)V - \sigma I_{1,p}\psi} + \frac{\sigma I_{1,p}(V-\psi)\left\{\left(I_{1,max} + \sigma I_{1,p}\right)V - \sigma I_{1,p}\psi\right\}}{2I_{1,max}}\right]}{\left[I_{dc,c} V + \sigma I_{dc,p}(V-\psi)\right]} \times 100 \tag{3.31}$$

where,

$$V = \frac{v_{in}}{V_{in,max}}$$

Since the load impedances of the both amplifiers decrease, the load lines of the carrier amplifier move up while maintaining the quasi-"$L$" curves, as shown in Fig. 3.27A, and the carrier amplifier operates in the saturation state. On the other hand, the current and voltage swings of the peaking amplifier increase in proportion to the drive level due to the proper load modulation, but the amplifier maintains the unsaturation state until it reaches to the minimum allowable output voltage level (point C). In this situation, the load lines of the amplifier are slightly elliptical until it reaches to the minimum allowable output voltage level. For the input power higher than that of the point C operation,

**Table 3.1** Operation State of the Carrier and Peaking Amplifiers According to Each Power Level

| | Low-Power Region (A) | Medium-Power Level (B) | High-Power Region (C) | High Power Level (D) |
|---|---|---|---|---|
| Carrier amplifier | Unsaturated state | *Saturation state* | *Saturation state* | *Saturation state* |
| Peaking amplifier | Turn off | Turn on | Unsaturated state | *Saturation state* |

the load line becomes the quasi-"*L*" curve, as shown in Fig. 3.27B. The peaking amplifier reaches to the saturation state after it passes the minimum allowable output voltage level, and both amplifiers have the half-sinusoidal current and maximally flat voltage waveforms at the maximum drive level (point D). The efficiency of the saturated Doherty amplifier, which is the average efficiency of the two amplifiers, is higher than the peak efficiency of the carrier amplifier due to the higher efficiency of the class C bias peaking amplifier, as shown in Fig. 3.27. The operation states of the Doherty amplifier are summarized in Table 3.1.

#### 3.4.2.2 Linearity of the Saturated Doherty Amplifier

The linearity of the saturated Doherty amplifier is worse than that of conventional Doherty amplifier because the carrier and peaking amplifiers are operated as saturated amplifiers to maximize the efficiency. In the low-power region, the linearity of the amplifier is entirely determined by the carrier amplifier like the conventional one. But the saturated amplifier has a poor linearity. Moreover, the carrier amplifier has a larger impedance load than that of the conventional one, and the amplifier reaches to saturation at a lower input drive level. Thus, the linearity is worse than the conventional Doherty amplifiers and is similar to the class F amplifier. However, in the high-power region, the linearity can be improved by the harmonic cancellation between the two amplifiers with appropriate gate biases like a conventional Doherty amplifier. Due to the linearity improvement mechanism, the Doherty amplifier can be more linear than the switching or saturated amplifiers.

Fig. 28A shows IM3 amplitudes of the carrier and peaking amplifiers and their phase difference. As already stated before, the carrier amplifier has relatively poor linearity with the high IM3 in the low-power region. In the high-power region where the carrier and peaking amplifiers are active, through proper gate bias arrangement of the two amplifiers, we can have IM3 amplitudes of both amplifiers are the same and the phase difference is maintained at $\sim 180°$. The IMD3 of the Doherty amplifier can be reduced by the harmonic cancellation, as shown in Fig. 3.28B. Therefore, the saturated Doherty amplifier can operate as a reasonably linear amplifier with very high efficiency in spite of the saturated operation, while the class F amplifier cannot.

### 3.4.3 Improved Harmonic Control Circuit for Saturated Amplifier

In the previous session, we have discussed the Doherty amplifier based on the class F amplifier. However, the saturated amplifier with the second harmonic control, which is not the short or open circuit but an inductive load, can provide very high efficiency,

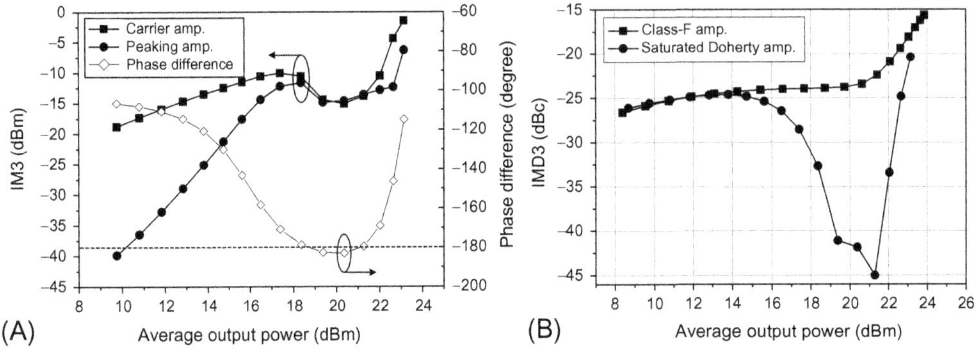


**Fig. 3.28** Two-tone simulated results of the saturated Doherty amplifier: (A) IM3 amplitudes of each amplifier and phase difference and (B) IMD3.

with very simple circuit and the Doherty amplifier based on the saturated amplifier is introduced in this section. The detailed operational behavior of the saturated amplifier can be found in Further Reading of this chapter. A conventional way to tune the second harmonic load in the saturated amplifier is to resonate at the second harmonic frequency using a capacitor. This capacitor is located at the end of the drain bias line and tunes the second harmonic impedance in the slightly inductive region from the resonation. Unfortunately, the second harmonic tuning is affected by the fundamental load modulation of the Doherty network, leading to a mismatch. Therefore, a new second harmonic tuning circuit, without being affected by the load modulation, is needed.

For the carrier amplifier, we generally employ $R_O = 50\,\Omega$ to $R_{OPT}$ matching circuit. To get the $2R_{OPT}$ load, two impedance transformers (50–25 and 25–100 Ω) and the offset line is used as shown in Fig. 3.29A. The impedance transformers have 90° electric length

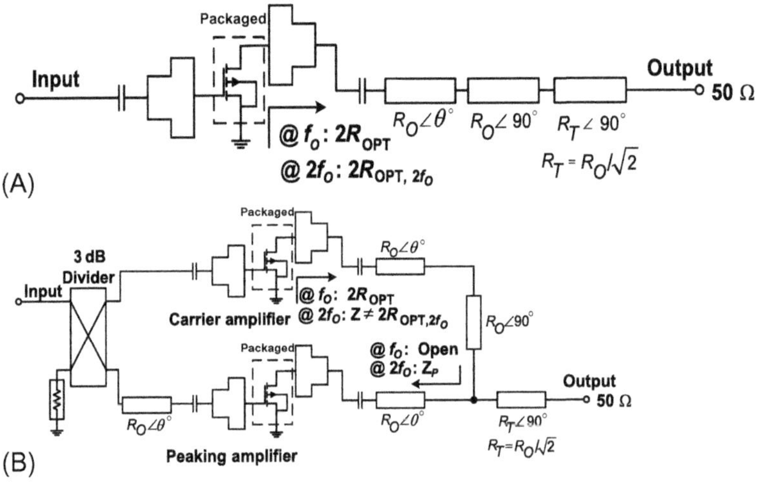


**Fig. 3.29** Second harmonic impedance mismatch: (A) carrier amplifier for $2R_{OPT}$ matching and (B) second harmonic mismatch of the carrier amplifier in the Doherty amplifier.

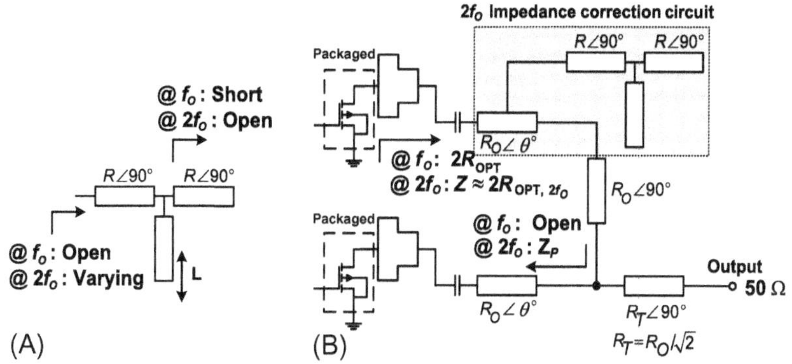


**Fig. 3.30** Second harmonic impedance correction circuit: (A) structure of the second harmonic impedance correction circuit and (B) implementation in the DPA structure.

at the fundamental frequency, and the second harmonic matching impedance is not affected by these transformers. Also, the offset line has 50Ω characteristic impedance. Therefore, the second harmonic matching impedance for the carrier amplifier is not affected by the load modulation circuit. But an offset line is employed at the peaking amplifier also to make an open circuit at the combining node. However, this offset line cannot provide the open impedance at the second harmonic frequency as shown in Fig. 3.29B, inducing the second harmonic impedance mismatch at the carrier amplifier.

To correct the second harmonic impedance mismatch of the carrier amplifier, a correction circuit is needed. The correction circuit should be independent of the fundamental load modulation and easy to utilize with other techniques required to improve the Doherty amplifier performance. With these constraints, the best location to place the correction circuit is on the offset line since the offset line functions only at the fundamental frequency. The second harmonic impedance correction circuit consists of an adjustable line between the two quarter-wave stubs at the fundamental frequency. The two quarter-wave lines form an open circuit at a connection point on the offset line, and the fundamental impedance is not affected by the correction circuit. But the second harmonic impedance can be controlled as shown in Fig. 3.30. Using this second harmonic control circuit, the Doherty amplifier based on the saturated amplifier can be realized.

## 3.5 AVERAGE POWER TRACKING OPERATION OF BASESTATION DOHERTY AMPLIFIER

The Doherty power amplifier installed in the base station is usually designed at one specific output power level. However, the total data usage can vary depending on the time of day. Thus, the output power of the power amplifier needs to be adjusted according to the data usage to reduce the system power consumption. As the average output power is decreased, a normal Doherty amplifier has a lower efficiency.

Therefore, the Doherty amplifier needs to be reconfigured properly for the variation of the output power. The simplest solution is an average power tracking (APT) of the Doherty power amplifier through its drain bias voltage control, similar to the APT of a handset power amplifier. The drain and gate bias voltages should be adjusted for the optimized operation at the different output power levels.

In a conventional Doherty amplifier design, it is biased at fixed gate and drain bias voltages as shown in Fig. 3.31A. However, to cover a wide dynamic range of the average output power with high efficiency, the Doherty amplifier needs to be reconfigured for the drain bias adaptation. The efficiency profile through the APT operation is depicted in Fig. 3.31B. As the peak output power is decreased from $P_1$ to $P_2$, the drain bias voltage is reduced to have high efficiency at the low peak power $P_2$. As the peak output power is

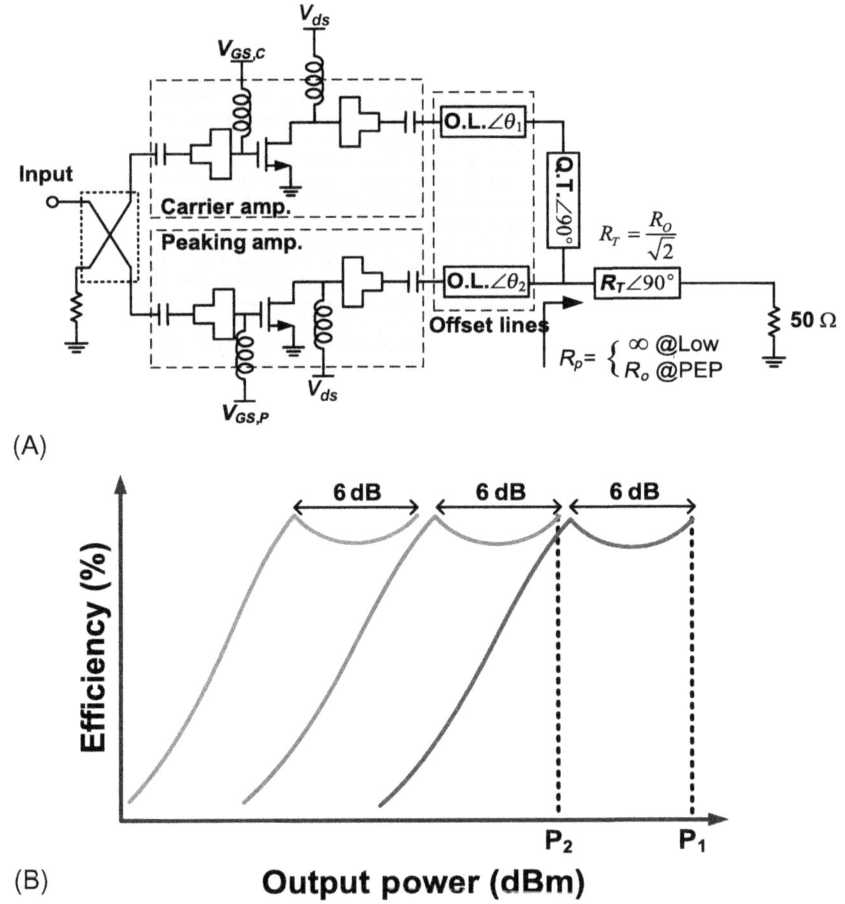


**Fig. 3.31** (A) Schematic of the conventional Doherty amplifier. (B) Efficiency profile of the average power tracking Doherty amplifier.

decreased, the first peak efficiency point is also decreased. The peaking amplifier should be turned on at the lower input power. Therefore, the gate bias voltage of the peaking amplifier should be adjusted accordingly. These behaviors are discussed in the following section.

### 3.5.1 Derivation of Drain Bias Control Voltage and Output Power

The optimum load lines of an amplifier, which is biased at class B with different drain biases, are shown in Fig. 3.32. When the peak output power $P_1$ is decreased to $P_2$, the optimum drain bias voltage needs to be reduced from $V_{ds,1}$ to $V_{ds,2}$, and the knee voltage is moved from $V_{k1}$ to $V_{k2}$. Therefore, the ratio of $P_{1,\text{linear}}$ and $P_{2,\text{linear}}$ can be derived as

$$\frac{p_{2,\text{linear}}}{p_{1,\text{linear}}} = \frac{V_{ds,2} - V_{k2}}{V_{ds,1} - V_{k1}} \cdot \frac{I_{\max,2}}{I_{\max,1}} \tag{3.32}$$

Because the load impedance is fixed, the ratios of the current and voltage swings are the same for the two power levels and are given by

$$\frac{V_{ds,2} - V_{k2}}{V_{ds,1} - V_{k1}} = \frac{I_{\max,2}}{I_{\max,1}} = \sqrt{\frac{p_{2,\text{linear}}}{p_{1,\text{linear}}}} \tag{3.33}$$

The knee voltages $V_{k1}$ and $V_{k2}$ can be described as

$$V_{k1} = R_{on} \cdot I_{\max,1} \quad \text{and} \quad V_{k2} = R_{\text{on}} \cdot I_{\max,2} \tag{3.34}$$

and the ratio of $V_{k1}$ and $V_{k2}$ can be expressed as

$$\frac{V_{k2}}{V_{k1}} = \frac{I_{\max,2}}{I_{\max,1}} = \sqrt{\frac{p_{2,\text{linear}}}{p_{1,\text{linear}}}} \tag{3.35}$$

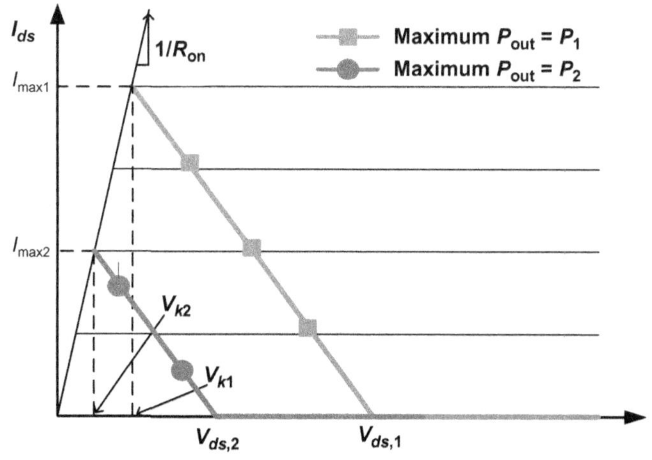


**Fig. 3.32** Load lines with nonzero knee voltages for the maximum powers of $P_1$ and $P_2$.

By substituting Eq. (3.35) into Eq. (3.33), the optimum drain bias voltage $V_{ds,2}$ can be derived as

$$V_{ds,2} = V_{ds,1} \cdot \sqrt{\frac{p_{2,\text{linear}}}{p_{1,\text{linear}}}} \tag{3.36}$$

As a result, the drain bias voltages for the carrier and peaking amplifiers should be reduced in proportion to the square root of the peak output power ratio.

### 3.5.2 Derivation for Gate Bias Control Voltage

As the peak output power is decreased from $P_1$ to $P_2$, the input power should be reduced linearly. Therefore, the peaking amplifier should be turned on at a 6 dB back-off level from the lower input power. To be turned on in the optimum back-off level, the gate bias voltage of the peaking amplifier needs to be increased linearly also. The input power levels, when the peaking amplifier is turned on, can be derived as

$$P_{1,\text{in}} = P_1 - 6 - G_1(\text{dBm}) \quad \text{for} \quad P_1\text{output} \tag{3.37}$$

$$P_{2,\text{in}} = P_2 - 6 - G_2(\text{dBm}) \quad \text{for} \quad P_2\text{output} \tag{3.38}$$

where $G_1$ and $G_2$ are gains for $P_1$ and $P_2$ power operations, respectively. However, the lower drain bias operation leads to a lower gain, and the turn-on voltage of the peaking amplifier should be adjusted following Eq. (3.38).

In Fig. 3.33A, the extracted input capacitance, $C_{\text{in}}$, is shown as a function of the drain bias voltage. As the drain bias voltage is decreased, the input capacitance is increased, and the gain of the carrier amplifier is decreased as shown in Fig. 3.33B. Therefore, the gate voltage of the peaking amplifier should be reduced accordingly. The two mechanisms compensate the variation of the gate bias voltage. As a result, the optimum gate bias voltage of the peaking amplifier is almost the same for the reconfigured Doherty amplifier for the APT operation.

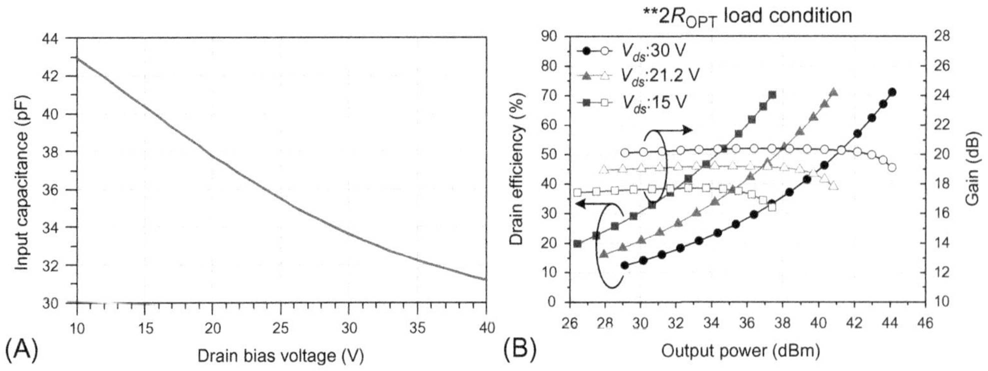


**Fig. 3.33** (A) Extracted input capacitance $C_{\text{in}}$ with the drain bias voltage variation. (B) Simulated CW characteristics of the carrier amplifiers with the drain bias voltage control.

### 3.5.3 Simulation Result of Reconfigured Doherty Amplifier for APT Operation

To verify the analysis for the bias voltage adaptations of the reconfigured Doherty amplifier, a Doherty amplifier is designed at 1.94 GHz using a Cree GaN HEMT CGH40045 model. The output matching circuits are optimized for efficiency at the drain bias voltage of 30 V. The harmonics are matched to the optimum impedances for an efficient saturated operation as described in the previous section. The simulation is conducted for the peak output powers with 3 dB step. As the peak output power is decreased to 47 and 44 dBm from 50 dBm, the drain bias voltage is reduced to 21.2 and 15 V, respectively. The gate bias voltage of the peaking amplifier is almost the same, slightly increased from −6 to −5.5 V for 47 dBm and −5.3 V for 44 dBm. The simulation results of the reconfigured Doherty amplifier are shown in Fig. 3.34. When the drain bias voltage is 30 V, the efficiencies are around 70% at the peak power (49.5 dBm) and 6 dB back-off level. As the drain bias voltage is reduced to 21.2 V and 15 V, the efficiencies are around 70% and 72%, respectively at the two power levels.

### 3.5.4 Implementation and Experimental Results

To validate the simulated results of the reconfigured Doherty amplifier for the APT operation, the amplifier is implemented. The Wilkinson power divider is designed for an even power split. The carrier amplifier is set to a deep class AB bias with an idle current of 180 mA. The experiment is conducted with the peak output powers of 50, 47, and 44 dBm, decreased in 3 dB step, and the bias conditions are the same with the simulated ones. The measured results for a CW signal are shown in Fig. 3.35. As the peak output powers are decreased with the 3 dB step, the drain bias voltages are reduced from 30 to

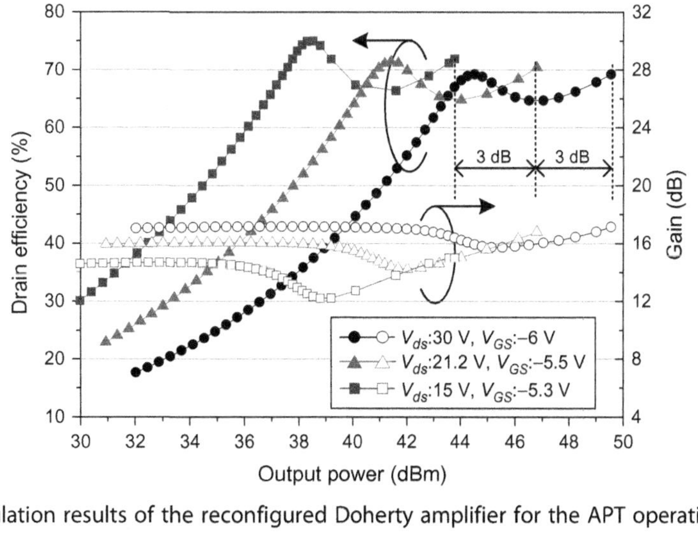


**Fig. 3.34** Simulation results of the reconfigured Doherty amplifier for the APT operation.

21.2 and 15 V, respectively. The gate bias voltage is remained almost the same, changing from −6 to −5.3 V as the drain bias voltage is decreased. As the result, the efficiency and gain characteristics of the reconfigured Doherty amplifier are moved by the 3 dB step as shown in the figure. The Doherty amplifier is tested using a 10 MHz LTE signal with a 6.5 dB PAPR at 1.94 GHz with a 6.5 dB PAPR. As shown in Fig. 3.36, the measured drain efficiencies and gains are 53.2%, 12.8 dB at 42.9 dBm; 54.3%, 11.2 dB at 39.9 dBm; and 53.4%, 9.1 dB at 37 dBm, respectively.

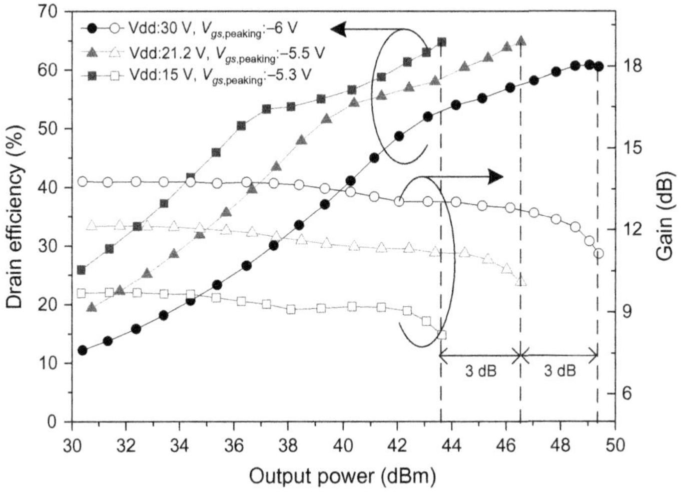


**Fig. 3.35** Measured results of the reconfigured Doherty amplifier for CW signal at 1.94 GHz.

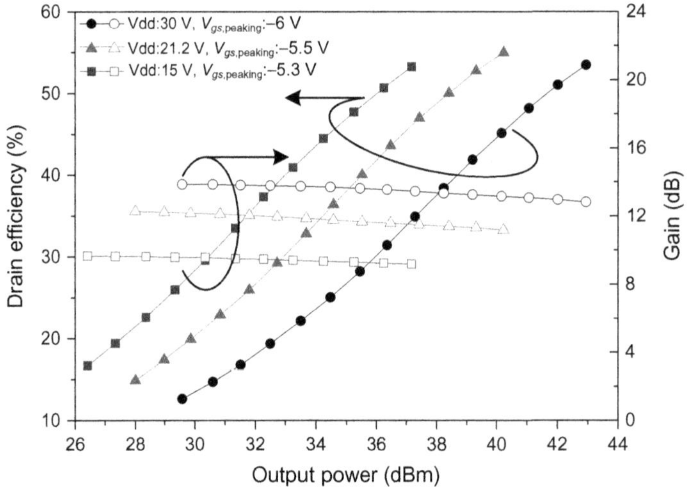


**Fig. 3.36** Measured results of the reconfigured Doherty amplifier for LTE signal at 1.94 GHz.

## FURTHER READING


[1] T. Yamamoto, et al., in: 50% drain efficiency Doherty amplifier with optimized power range for W-CDMA signal, IEEE MTT-S International Microwave Symposium Digest, June 2007, pp. 1263–1266.

[2] J. Moon, et al., Efficiency enhancement of Doherty amplifier by mitigating the knee voltage effect, IEEE Trans. Microwave Theory Tech. 59 (1) (2011) 143–152.

[3] Y. Park, et al., GaN HEMT MMIC Doherty power amplifier with high gain and high PAE, IEEE Microwave Wireless Compon. Lett. 25 (3) (2015) 187–189.

[4] S. Kim, et al., in: Optimized peaking amplifier of Doherty amplifier using an inductive input second harmonic load, Proc. 46th Eur. Microw. Conf., London, United Kingdom, October 2016, pp. 3–7.

[5] J. Kim, et al., Analysis of a fully matched saturated Doherty amplifier with excellent efficiency, IEEE Trans. Microwave Theory Tech. 56 (2) (2008) 328–338.

[6] J. Kim, et al., Saturated power amplifier optimized for efficiency using self-generated harmonic current and voltage, IEEE Trans. Microwave Theory Tech. 59 (8) (2011) 2049–2058.

[7] S.C. Cripps, RF Power Amplifiers for Wireless Communications, Artech House, Norwood, MA, 2006.

[8] Y. Park, et al., Analysis of average power tracking Doherty power amplifier, IEEE Microwave Wireless Compon. Lett. 25 (7) (2015) 481–483.

[9] A. Mohamed, et al., Doherty power amplifier with enhanced efficiency at extended operating average power levels, IEEE Trans. Microwave Theory Tech. 61 (12) (2013) 4179–4187.

[10] Y. Park, et al., A highly efficient power amplifier at 5.8 GHz using independent harmonic control, IEEE Microwave Wireless Compon. Lett. 27 (1) (2017) 76–78.

CHAPTER FOUR

# Advanced Architecture of Doherty Amplifier

So far, we have considered the Doherty amplifier based on two identical amplifiers. In this case, the first peak-efficiency point is located at the fixed 6 dB back-off power. However, according to the modulated signal profile, the location of the peak point should be changed to get a higher efficiency for amplification of the signal, or the efficiency can be even higher by increasing the number of the maximum efficiency points. These behaviors can be achieved by increasing the number of the unit amplifiers, one carrier amplifier and $(N-1)$ peaking amplifiers. When the $(N-1)$ unit amplifier cells are used for the peaking amplifier with one load-modulation circuit, which is called $N$-way Doherty amplifier, the back-off level for the first peak-efficiency point is increased accordingly. When the $(N-1)$ unit amplifier cells are used for the peaking amplifier and the cells modulate the load sequentially using the separate load-modulation circuits, the number of the peak-efficiency points is increased accordingly and is called $N$-state Doherty amplifier. These amplifiers can deliver a higher efficiency for amplification of a highly modulated signal. The $N$-way Doherty amplifier can be used as a highly linear amplifier by tuning the multiple cells for accurate harmonic cancellation. The design concepts of these amplifiers are described in this chapter.

## 4.1 *N*-WAY DOHERTY AMPLIFIER

The two-way Doherty amplifier, the basic architecture of the Doherty amplifier we have discussed so far, has two identical unit amplifiers for the carrier amplifier and peaking amplifier and has two maximum efficiency points at the 6 dB back-off output power and the peak output power. The multiway Doherty amplifier consists of $N$ identical unit cells, one cell for the carrier amplifier and $(N-1)$ cells for the peaking amplifier, and is called a multi (in this case $N$)-way Doherty amplifier. The total size of the peaking amplifier $(1 \sim N-1)$ compared with that of the carrier amplifier determines the back-off output power level for the maximum efficiency point. The efficiency between the two peak-efficiency points, however, drops further as the separation between the two points increases. This architecture is useful for adjusting the first peak-efficiency power point closer to the average power region, which is required for efficient amplification of a signal with a high peak-to-average power ratio (PAPR).

*Doherty Power Amplifiers*
https://doi.org/10.1016/B978-0-12-809867-7.00004-1

### 4.1.1 Load Modulation of *N*-way Doherty Amplifier

Fig. 4.1 shows an equivalent circuit model of the *N*-way Doherty amplifier with *N* identical ideal current sources. $I_c$, $I_c'$, and $I_p$ represent the drain current of the carrier amplifier, the current after the inverter, and the drain current of the unit peaking cell, respectively. Since the inverter is lossless, a constant power flows through the inverter. Therefore, $I_c'$ can be calculated from the constant power flow concept:

$$I_c' = I_c \cdot \sqrt{\frac{Z_c}{Z_c'}} = I_c \cdot \frac{R_0}{Z_c'} = I_c \cdot \frac{Z_c}{R_0} \tag{4.1}$$

The load modulation is carried out by the current ratio between the carrier amplifier and peaking amplifier, and the load impedances of the unit cells can be derived by the load-modulation principle described in Chapter One:

$$V_0 = \frac{R_0}{N} \cdot \left(I_c' + I_{p,\mathrm{Total}}\right) = \frac{R_0}{N} \cdot \left(I_c' + \sum_1^{N-1} I_p\right) \tag{4.2}$$

$$Z_c' = \frac{V_0}{I_c'} = \frac{R_0}{N}\left(1 + \frac{\sum_1^{N-1} I_p}{I_c'}\right) = \frac{R_0}{\left(N - \frac{\sum_1^{N-1} I_p}{I_c}\right)} \tag{4.3}$$

$$Z_c = \frac{{R_0}^2}{Z_c'} = R_0\left(N - \frac{\sum_1^{N-1} I_p}{I_c}\right) \tag{4.4}$$

$$Z_P = \frac{V_0}{I_p} = \frac{R_0}{NI_p} \cdot \left(I_c' + \sum_1^{N-1} I_p\right) = R_0 \cdot \frac{I_c}{\sqrt{I_p\left(NI_c - \sum_1^{N-1} I_p\right)}} \tag{4.5}$$

In Eqs. (4.4), (4.5), $Z_c$ and $Z_p$ are the load impedances of the carrier cell and peaking unit cell, respectively.

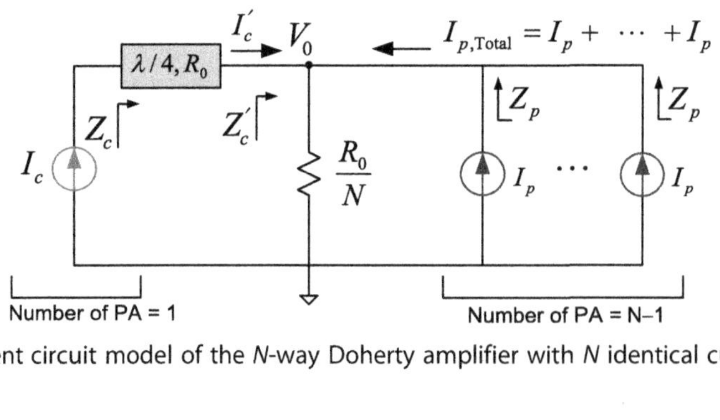


**Fig. 4.1** Equivalent circuit model of the *N*-way Doherty amplifier with *N* identical current sources.

If the current ratio of the peaking cell to the carrier cell is α, the load impedances of the unit cells, for a given input drive $V_{\text{in}}$, can be written as follows:

$$\alpha(V_{\text{in}}) = \frac{I_p(V_{\text{in}})}{I_c(V_{\text{in}})} \tag{4.6}$$

$$Z_c = R_0 \cdot [N - \alpha(V_{\text{in}})(N-1)] \tag{4.7}$$

$$Z_P = \frac{R_0}{\sqrt{\alpha(V_{\text{in}}) \cdot [N - (N-1)\alpha(V_{\text{in}})]}} \tag{4.8}$$

Since the unit cells are identical, $\alpha$ should be changed from 0 to 1 according to the input power level. The peaking amplifier should be turned on at the input voltage of $V_{\text{max}}/N$, where $V_{\text{max}}$ is the maximum input voltage swing and the transconductance of the peaking cell should be larger than that of the carrier amplifier by $1+1/N$. Because of the early turn-on of the $N$-way peaking amplifier, the uneven drive problem is reduced compared with the standard Doherty amplifier.

Accordingly, the load impedances are modulated with $\alpha(V_{\text{in}})$ as follows:

$$Z_c = \begin{cases} N \cdot R_0, & \text{at } V_{\text{in}} = \dfrac{V_{\text{max}}}{N} \text{ with } \alpha(V_{\text{in}}) = 0 \\ R_0, & \text{at } V_{\text{in}} = V_{\text{max}} \text{ with } \alpha(V_{\text{in}}) = 1 \end{cases} \tag{4.9}$$

$$Z_P = \begin{cases} \infty, & \text{at } V_{\text{in}} = \dfrac{V_{\text{max}}}{N} \text{ with } \alpha(V_{\text{in}}) = 0 \\ R_0, & \text{at } V_{\text{in}} = V_{\text{max}} \text{ with } \alpha(V_{\text{in}}) = 1 \end{cases} \tag{4.10}$$

Since all the peaking cells are modulated in the same way, the total impedance of the peaking amplifier is reduced by $1/(N-1)$ due to the parallel connection of the cells. The impedance of the carrier amplifier is increased in proportion to $N$ when the peaking amplifier is turned off. At the peak power, all the cells see the same load impedance of $R_0$.

Fig. 4.2 shows the load-modulation behaviors of the two-way Doherty amplifier and three-way Doherty amplifier. As shown in the above analysis, the peaking cells of the three-way Doherty amplifier are turned on at one-third of the maximum input voltage swing, and the load impedance of the cells is modulated from open to $R_0$. The load impedance of the carrier amplifier is modulated from $3R_0$ to $R_0$ as the peaking amplifier generates the current from zero to the maximum value. It should be remembered that the peaking amplifier has $(N-1)$ unit cells and the total current of the peaking amplifier is $(N-1)$ times larger than $I_{\text{max}}$.

The load lines of the carrier and peaking amplifiers are illustrated in Fig. 4.3. Before the peaking amplifier turns on, the carrier amplifier operates as a single class B amplifier. When the carrier amplifier reaches to the maximum efficiency point, the peaking amplifier is turned on. After turned on, the load impedance of the carrier amplifier is modulated, and the load line always follows the saturated region, generating the maximum

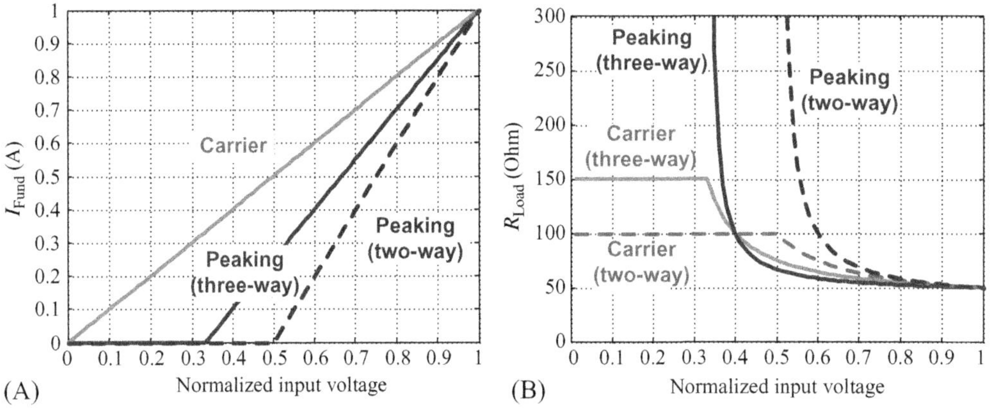


(A) (B)

**Fig. 4.2** Load modulation of *N*-way Doherty amplifier: (A) drain current and (B) load impedances of the unit cells.

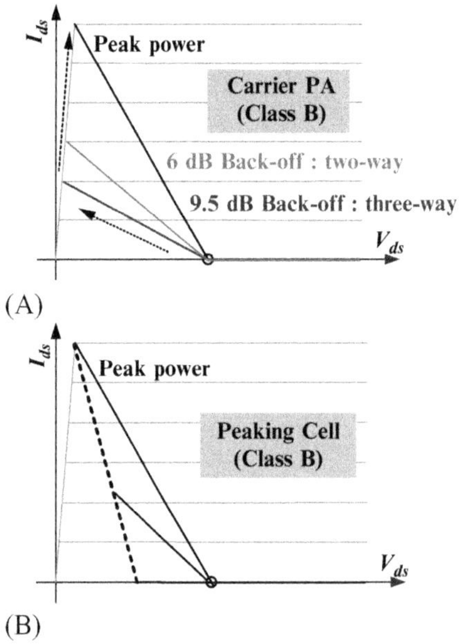


**Fig. 4.3** Load lines of the *N*-way Doherty amplifier: (A) carrier cell and (B) peaking cell.

efficiency as explained in Chapter One. The load impedance of the peaking amplifier is also modulated, but the amplifier delivers the peak efficiency only at the maximum output power point. For the load lines, it is assumed that transconductance of the peaking cell is $N/(N-1)$ times larger than that of the carrier amplifier. Therefore, the amplifiers reach the maximum current level simultaneously although the peaking amplifier is biased lower. It can be achieved using a bigger-size device for the peaking cells, but in real design, the same-size devices are employed to fully utilize the precious device resource, resulting in deviation from the ideal behaviors. This problem can be solved by the uneven drive or gate-bias adaptation, which is discussed in Chapter Two.

### 4.1.2 Efficiency of *N*-Way Doherty Amplifier

Using Eqs. (4.7), (4.8), the efficiency variation of the Doherty amplifier with the input power level can be derived. For the analysis, the transconductances of the devices are assumed to be uniform, generating the fundamental current components linearly with the input voltage as shown in Fig. 4.4. The carrier amplifier and peaking amplifier are operated at class B mode with the different turn-on voltages. The currents of the two amplifiers are given by

$$I_c(V_{in}) = \frac{V_{in}}{V_{max}} \cdot I_{max} \tag{4.11}$$

$$I_p(V_{in}) = \begin{cases} 0, & \text{for } 0 < V_{in} < \dfrac{V_{max}}{N} \\ I_{max} \cdot \left( \dfrac{N \cdot V_{in}}{V_{max}} - 1 \right), & \text{for } \dfrac{V_{max}}{N} < V_{in} < V_{max} \end{cases} \tag{4.12}$$

Here, $I_{max}$ is the maximum current of the single cell.

For the input voltage level of $0 \leq v_{in} < \frac{V_{max}}{N}$, the RF and DC powers can be calculated as follows:

$$\begin{aligned} P_{RF}(v_{in}) &= \frac{1}{8} \cdot I_C(v_{in})^2 \cdot Z_C(v_{in}) = \frac{I_{max}{}^2}{8} \cdot \left( \frac{v_{in}}{V_{max}} \right)^2 \cdot R_0 \cdot [N - \alpha(V_{in})(N-1)] \\ &= \frac{N \cdot I_{max} \cdot V_{dc}}{4} \cdot \left( \frac{v_{in}}{V_{max}} \right)^2 \left( \text{where, } \alpha(v_{in}) = 0 \text{ for } 0 \leq v_{in} < \frac{V_{max}}{N} \right) \end{aligned} \tag{4.13}$$

$$P_{DC}(v_{in}) = V_{dc} \cdot I_{dc}(v_{in}) = V_{dc} \cdot \frac{1}{\pi} \cdot I_{max} \cdot \left( \frac{v_{in}}{V_{max}} \right) \tag{4.14}$$

At the input level of $V_{max}/N \leq v_{in} < V_{max}$, the RF and DC powers are given by

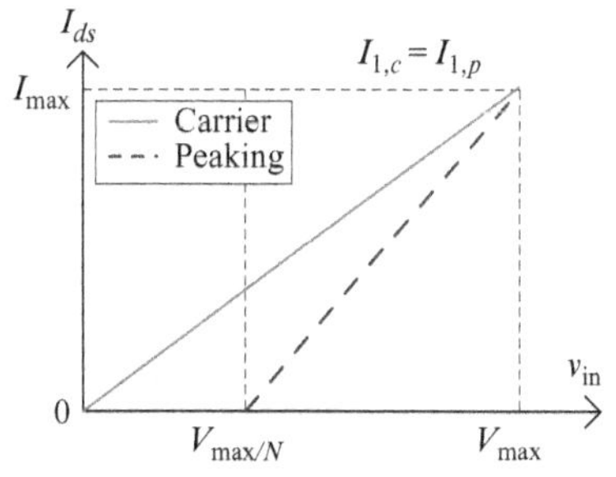


**Fig. 4.4** The current profiles of the carrier and peaking cells.

$$\begin{aligned}
P_{RF}(v_{\text{in}}) &= \frac{1}{8}\cdot I_C(v_{\text{in}})^2\cdot Z_C(v_{\text{in}}) + \frac{1}{8}\cdot I_P(v_{\text{in}})^2\cdot Z_P(v_{\text{in}}) \\
&= \frac{I_{\max}{}^2}{8}\cdot\left(\frac{v_{\text{in}}}{V_{\max}}\right)^2\cdot R_0\cdot[N-\alpha(V_{\text{in}})(N-1)] + \frac{I_{\max}{}^2}{8}\cdot\left(\frac{V_{\text{in}}}{V_{\max}}\cdot N-1\right)^2\cdot\frac{R_0}{\sqrt{\alpha(V_{\text{in}})\cdot[N-(N-1)\alpha(V_{\text{in}})]}} \\
&= \frac{1}{4}\cdot I_{\max}\cdot V_{dc}\cdot\left[\left(\frac{v_{\text{in}}}{V_{\max}}\right)^2\cdot[N-\alpha(V_{\text{in}})(N-1)] + \left(\frac{V_{\text{in}}}{V_{\max}}\cdot N-1\right)^2\cdot\frac{1}{\sqrt{\alpha(V_{\text{in}})\cdot[N-(N-1)\alpha(V_{\text{in}})]}}\right]
\end{aligned} \tag{4.15}$$

$$\begin{aligned}
P_{DC}(v_{\text{in}}) &= V_{dc}\cdot\left[I_{dc}(v_{\text{in}}) + (N-1)\cdot I_{dp}(v_{\text{in}})\right] \\
&= \frac{1}{\pi}\cdot I_{\max}\cdot V_{dc}\cdot\left[\frac{v_{\text{in}}}{V_{\max}} + \left(\frac{V_{\text{in}}}{V_{\max}}\cdot N-1\right)\right]
\end{aligned} \tag{4.16}$$

The efficiency can be calculated from Eqs. (4.13) to (4.16).

$$\eta = \begin{cases}
\dfrac{\pi}{4}\cdot\dfrac{N\cdot v_{\text{in}}}{V_{\max}}, & 0 \le v_{\text{in}} < \dfrac{V_{\max}}{N} \\[2ex]
\dfrac{\pi}{4}\cdot\dfrac{\left[\left(\dfrac{v_{\text{in}}}{V_{\max}}\right)^2\cdot[N-\alpha(V_{\text{in}})(N-1)] + \left(\dfrac{V_{\text{in}}}{V_{\max}}\cdot N-1\right)^2\cdot\dfrac{1}{\sqrt{\alpha(V_{\text{in}})\cdot[N-(N-1)\alpha(V_{\text{in}})]}}\right]}{\dfrac{v_{\text{in}}}{V_{\max}} + \left(\dfrac{V_{\text{in}}}{V_{\max}}\cdot N-1\right)} & \dfrac{V_{\max}}{N} \le v_{\text{in}} < V_{\max} \\[2ex]
\dfrac{\pi}{4} & v_{\text{in}} = V_{\max}
\end{cases} \tag{4.17}$$

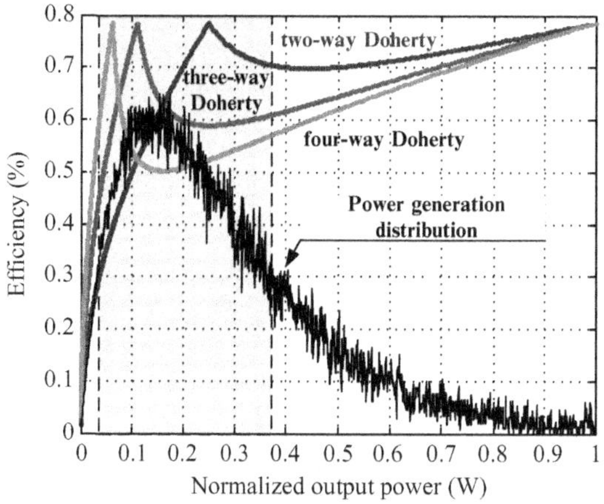


**Fig. 4.5** Simulated efficiencies of *N*-way Doherty amplifiers versus the output power level.

Fig. 4.5 shows the calculated efficiency of the *N*-way Doherty amplifier versus the output power level calculated by Eq. (4.17). The maximum efficiency of 78.5% ($\pi/4$) is obtained at the input power levels of "$V_{max}/N$" and "$V_{max}$." The output power level for the first peak-efficiency point is $1/N^2$ or 20·log ($1/N$) dB down from the peak output power. The back-off power level for the maximum efficiency is increased as the size of the peaking amplifier (number of peaking cells) is increased. The three-way and four-way Doherty amplifiers deliver the maximum efficiencies at the back-off powers of 9.54 dB (1/9 of normalized output power) and 12 dB (1/16 of normalized output power), respectively. The efficiency between the two power points is decreased as the separation is increased.

To evaluate the efficiency of the Doherty amplifier for amplification of a modulated signal, the 802.16e Mobile WiMAX signal with 8.5 dB PAPR is used. The drain efficiency (DE) of *N*-way Doherty amplifier for the modulated signal is determined by following equation:

$$DE = \frac{\int \text{prob.}(v_{\text{in}}) \cdot P_{\text{out}}(v_{\text{in}}) dv_{\text{in}}}{\int \text{prob.}(v_{\text{in}}) \cdot P_{dc}(v_{\text{in}}) dv_{\text{in}}} \tag{4.18}$$

The prob.$(v_{\text{in}})$ is the occurrence probability of the $v_{\text{in}}$ in the modulated input signal. The overall DE is determined by ratio of the multiplications of the probability distribution and power generation terms ($P_{\text{out}}$) over the multiplications of the distribution and DC power ($P_{dc}$). The numerator in Eq. (4.18) is named as the power generation distribution (PGD) of the signal, which represents the real power generation at the input level of the modulated signal. The normalized power generation amount is also shown in Fig. 4.5. The average power generation point is 7.5 dB lower than the peak power for 802.16e mobile WiMAX signal with 8.5 dB PAPR. The distribution indicates that the dominant power generation region for amplification of the modulated signal is 0.03 ~ 0.37, indicating that three-way Doherty amplifier is the best suited one. According to the PAPR of the modulated signal, the important power generation region indicated by the PGD is changed, and the *N*-way Doherty amplifier should be chosen properly according to the distribution. The calculated DEs for the *N*-way Doherty amplifiers are summarized in Table 4.1.

The calculated DEs for the signal with various PAPRs are illustrated in Fig. 4.6. Because the back-off level of PGD is lower than that of PAPR, the two-way Doherty amplifier delivers the highest efficiency for the signal with PAPR of up to 7.5 dB, and the three-way is better for the higher PAPR signals.

The *N*-way Doherty amplifier is not the optimum architecture for an efficient transmitter because it does not maintain the peak efficiency at the important PGD region. These low-efficiency regions are originated from the unsaturated operations of the carrier amplifier at the low-power region and the peaking amplifier at the high-power region. To reduce the efficiency degradation, the high-efficiency characteristic should be maintained over the important PGD region. It can be achieved by increasing the number of the maximum efficiency points by increasing the load-modulation networks (*N*), and the *N*-stage Doherty amplifier has been suggested. At the next section, we will explore the *N*-stage Doherty amplifier for the highly efficient amplification of a signal with a high PAPR.

**Table 4.1** Calculated Drain Efficiencies of the *N*-Way Doherty Amplifiers for the 802.16e Mobile WiMAX Signal With 8.5 dB PAPR

| | Two-Way | Three-Way | Four-Way |
|---|---|---|---|
| $\eta_{avg}$ (%) | 59.5 | 62 | 56.5 |

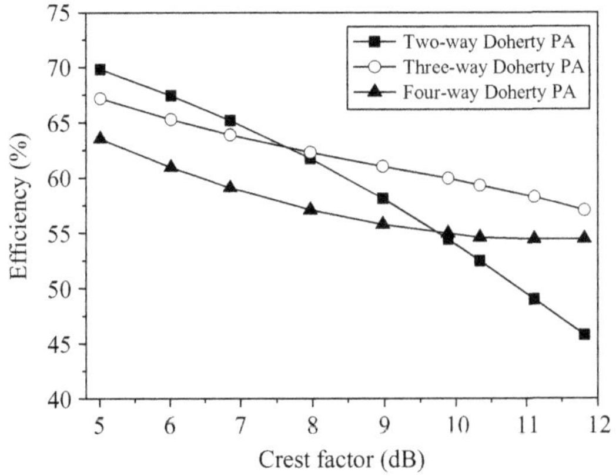


**Fig. 4.6** Calculated efficiency of *N*-way Doherty amplifier for the modulated signal with various PAPRs.

### 4.1.3 Linearity of *N*-way Doherty Amplifier

As discussed in Chapter One, the Doherty amplifier is inherently a linear power amplifier if the internal distortions generated by the carrier and peaking amplifiers are not considered. To realize a linear Doherty amplifier, the distortions should be minimized. For a linear Doherty amplifier, the carrier amplifier is biased at class AB and the peaking amplifier at class C. The class AB carrier amplifier should operate linearly by itself at the low power before the peaking amplifier turns on. At near or above the first peak-efficiency point, the carrier amplifier operates at the near-saturation mode, and a large distortion is generated from the carrier amplifier. The distortion at the saturated operation should be canceled by the distortion generated by the class C biased peaking amplifier. It is possible since the class AB mode carrier amplifier has a gain compression characteristic while the class C mode peaking amplifier has a large gain-expansion characteristic. By adjusting the two gain characteristics through the proper gate biases of the two amplifiers, a constant gain from the Doherty amplifier can be obtained. It is equivalent to cancellation of the harmonic distortions generated by the two amplifiers. The asymmetrical output powers from the carrier and peaking amplifiers with the adjusted biases are combined by the Doherty network, which is a very nice unique characteristic of Doherty amplifier.

The nonlinear output current of an active device can be expressed using Taylor series expansion by

$$I_{out} = gm_1 \cdot v_{in} + gm_2 \cdot v_{in}^2 + gm_3 \cdot v_{in}^3 + gm_4 \cdot v_{in}^4 + gm_5 \cdot v_{in}^5 + \cdots \tag{4.19}$$

where $v_{in}$ is the input gate voltage and $gm_{X's}$ are the $X$th-order expansion coefficients of the nonlinear $gm$. The third-order intermodulation IM3 current is mainly generated by $gm_3 \cdot v_{in}^3$ term and the fifth-order intermodulation IM5 current by $gm_5 \cdot v_{in}^5$ term in Eq. (4.19). For a sinewave input to Eq. (4.19), the fundamental component $I_{out_{fund}}$ is given by

$$I_{out_{fund}} = \left[ gm_1 \cdot v_{in} + \frac{9}{4} gm_3 \cdot v_{in}^3 + \frac{25}{4} gm_5 \cdot v_{in}^5 + \cdots \right] \tag{4.20}$$

Fig. 4.7 presents $gm_3$ curve according to the gate-bias voltage of a transistor. It is important to notice that $gm_3$ is positive for the gate bias close to the pinch off and is negative for the higher bias voltage. Because of the positive $g_{m3}$, the class C amplifier has a gain-expansion characteristic since this IM3 can generate the fundamental term as shown in Eq. (4.20), while the class AB amplifier with the negative $gm_3$ has a gain compression characteristic.

As mentioned earlier, for a linear Doherty amplifier, the carrier amplifier is biased at a class AB mode and should operate linearly by itself. The bias voltage of the peaking amplifier should be in class C mode in order to have proper positive $gm_3$ necessary for canceling the IMD3 current generated by the class AB carrier amplifier at a high power

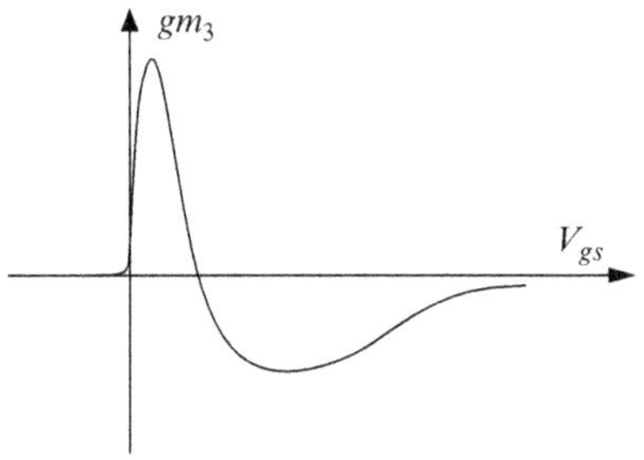


**Fig. 4.7** The $gm_3$ curve versus gate-bias voltage for an FET.

operation with near saturation. This class C peaking amplifier generates a considerable amount of the higher-order intermodulation terms, especially the fifth-order terms. However, a linear two-way Doherty amplifier can be realized by using this harmonic cancellation mechanism, and we will discuss this amplifier in Chapter Five "Linear Doherty Amplifier for Handset Application"

For the three-way case, which has two peaking cells, an extremely linear Doherty amplifier can be realized. For the amplifier, the required bias voltage for the proper IMD3 cancellation is higher than the two-way Doherty amplifier. Therefore, the peaking amplifiers operate more linearly without generating the excessive higher-order distortion terms, helping for linear operation. The gate bias of the higher-way Doherty amplifier should be even higher, but the higher-bias operation more than the three-way Doherty amplifier increases the sensitivity for the harmonic cancellation. Consequently, the best efficiency versus linearity characteristics should be considered.

The two-, three-, and four-way Doherty amplifiers are compared for their performances on linearity and efficiency. The basic amplifier cells have been designed using Motorola's 4 W PEP silicon LDMOSFET. The linearity is optimized by adjusting the gate biases of the peaking cells. The uneven powers from the two amplifiers are automatically combined by the load-modulation circuit. The class AB amplifiers for the reference points are realized by equally AB biasing all the unit cells of the Doherty amplifier. Fig. 4.8A shows the measured ACLRs of the two-way Doherty amplifier and class AB amplifier at 2.5 MHz offset and 5 MHz offset for one- and two-carrier downlink WCDMA signals (5 MHz bandwidth), respectively. For the one-carrier WCDMA signal, ACLR is improved by 4.69 dB at an output power of 27 dBm. Fig. 4.8B shows the measured PAEs of the two-way Doherty and class AB amplifiers for the one- and two-carrier downlink WCDMA signals. The PAE is improved by 6.45%, from 21.45% to 27.9% at the output power of 32 dBm. The optimized quiescent current of the peaking amplifier is 0.1 mA, while those of the carrier amplifier and the class AB amplifiers are 60 mA.

Fig. 4.9A shows the measured ACLRs of the three-way Doherty and class AB amplifiers. For the one-carrier WCDMA signal, ACLR is improved by 9.97 dB at the output power of 30 dBm. Fig. 4.9B shows the measured PAEs of the three-way Doherty and

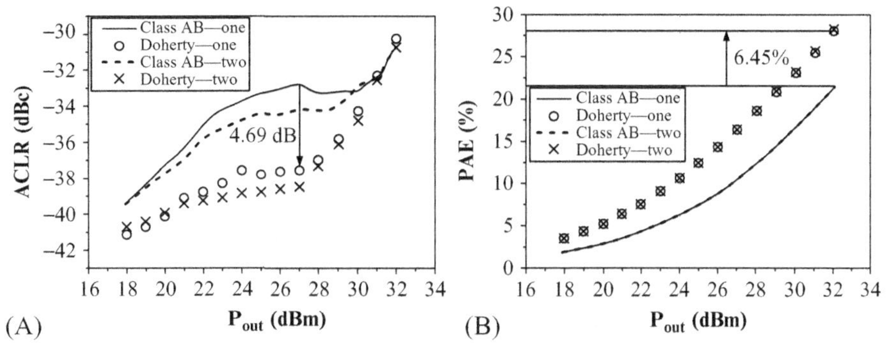


**Fig. 4.8** Measured performances of the two-way Doherty and class AB amplifiers with one- and two-carrier downlink WCDMA signals: (A) ACLRs and (B) PAEs.

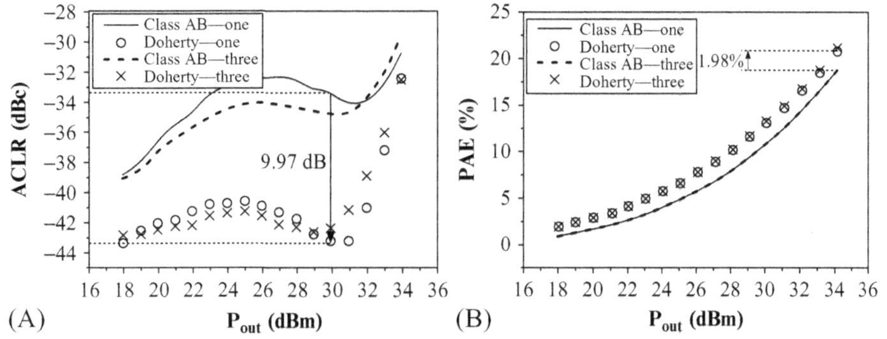


**Fig. 4.9** Measured performances of the three-way Doherty and class AB amplifiers with one- and two-carrier downlink WCDMA signals: (A) ACLRs and (B) PAEs.

class AB amplifiers. The PAE is improved slightly by 1.98% at an output power of 34 dBm.

Fig. 4.10A shows the measured ACLRs of the four-way Doherty and class AB amplifiers. For the one-carrier WCDMA signal, ACLR is improved by 8.83 dB at an output power of 32 dBm. Fig. 4.10B shows the measured PAEs of the four-way Doherty and class AB amplifiers. The PAE is improved by 2.5% at output power of 35 dBm. The optimized quiescent currents of the peaking amplifiers are 12.8 and 15.73 mA for the three- and four-way cases, respectively.

As shown, the three- or four-way Doherty amplifiers deliver greatly improved linearity over the two-way Doherty amplifier. However, for the four-way case, the linearity improvement is observed over a relatively narrow output power range compared with the three-way case, which means that the excessive numbers of the peaking cells make the IM3 cancellation very sensitive. Also, as the number of peaking cells is increased, the bias current is also raised for the intermodulation cancellation. Therefore, the

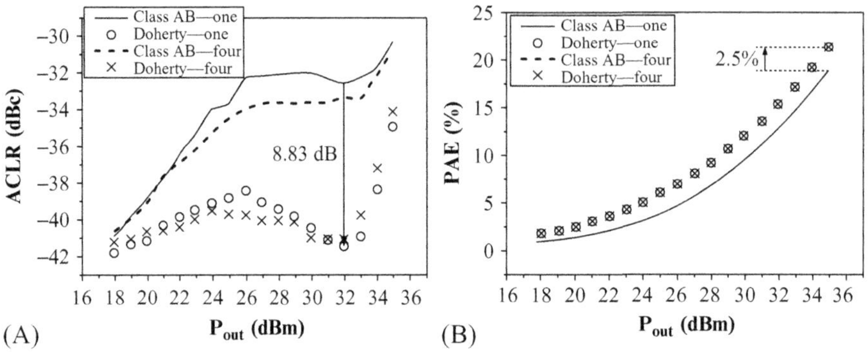


**Fig. 4.10** Measured performances of the four-way Doherty and class AB amplifiers with one- and two-carrier downlink WCDMA signals: (A) ACLRs and (B) PAEs.

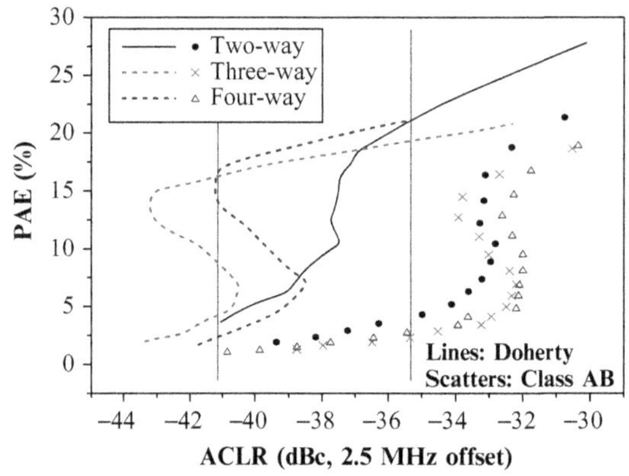


**Fig. 4.11** Efficiency versus linearity characteristics for the *N*-way Doherty and *N*-way class AB amplifiers.

load-modulation effect is reduced, and the efficiency is reduced, comparable with the class AB operation. Fig. 4.11 compares the data for the efficiency and linearity characteristics of the amplifiers. Compared with the normal class AB operated amplifiers, the *N*-way Doherty amplifiers deliver much more improved efficiency versus linearity. The two-way shows the best efficiency above −35 dBc of ACLR, while the three-way shows best efficiency below −41 dBc. For the intermediate region, the four-way shows slightly better efficiency than that of the three-way.

As shown, the three-way Doherty amplifier can be very linear, and the linearity can be further enhanced. To maximize the linearity of the amplifier, the carrier amplifier is designed to be very linear by biasing at near class A mode, and the biases of the peaking cells are adjusted for the IMD3 and IMD5 cancellations. The three-way Doherty amplifier has been implemented using three Motorola's MRF21060 (60 W PEP) LDMOSFETs at 2.14 GHz. In the experiments, the quiescent drain currents of the carrier and peaking amplifiers are set to 700 mA at $V_{DD} = 28$ V for the class AB amplifier operation. For the three-way Doherty amplifier, the quiescent drain current of the carrier amplifier is set to 820 mA at $V_{DD} = 28$ V, but the peaking cells are biased at class AB mode, around 260, 280 mA, respectively with $V_{DD} = 28$ V. Fig. 4.12 shows the measured ACLRs and PAEs of the class AB amplifier and three-way Doherty amplifier. Fig. 4.12A is for one-carrier WCDMA signal with 10 MHz bandwidth, and Fig. 4.12B is for two-carrier WCDMA signal with 10 MHz spacing. The three-way Doherty is very linear with ACLR of under 50 dBc. For the one-carrier WCDMA signal, ACLR is improved by about 10 dB compared with the class AB amplifier. The PAE is improved slightly by about 2% at the same output power. For the two-carrier WCDMA signal, ACLR of the three-way Doherty is improved by 6.8 dB at an output power 40 dBm, and the improvement of the PAEs is similar to the one-carrier case. Fig. 4.13 explains the linearity boosting mechanism of the three-way Doherty. For the three-way Doherty amplifier, the $IMD_3$ and $IMD_5$ are reduced simultaneously compared with the class AB case.

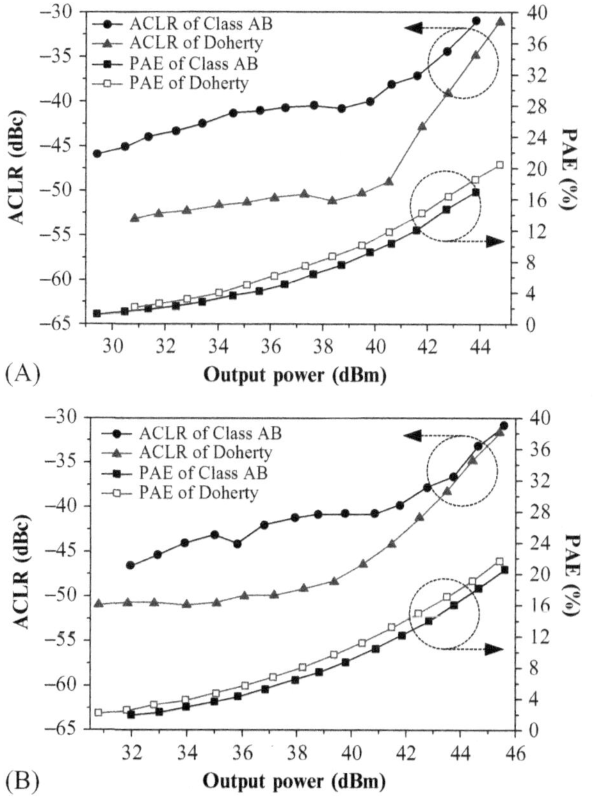


**Fig. 4.12** Measured ACLRs and PAEs of class AB, the three-way Doherty: (A) one-carrier downlink WCDMA signal and (B) two-carrier downlink WCDMA signal.

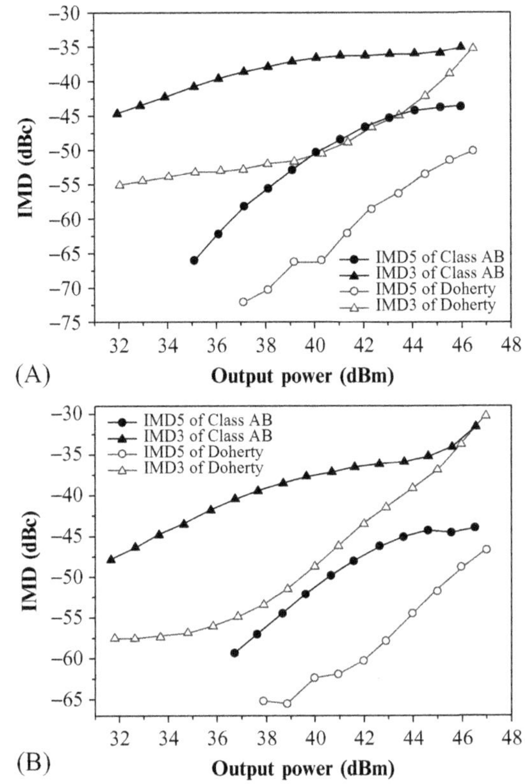


**Fig. 4.13** Measured $IMD_3$s and $IMD_5$s of class AB, the three-way Doherty: (A) 5 MHz spacing and (B) 10 MHz spacing.

In summary, the gate-bias voltages of the two cells of the peaking amplifier can be adjusted to cancel the IMD3 more accurately, and the peaking amplifier itself operates more linearly due to the higher gate bias. In this operation, the efficiency can be relatively high due to the near-normal Doherty operation. For the very linear operation, the three and fifth harmonics should be canceled using the two peaking cells. Since the fifth harmonic level is around 40–45 dBc, without the fifth harmonic cancel, the linearity cannot reach below −50 dBc. In this operation, the amplifier is very linear, but the efficiency is not high.

## 4.2 THREE-STAGE DOHERTY AMPLIFIER

The *N*-way Doherty amplifier creates two maximum efficiency points with different back-off power level using different-size devices for the carrier and peaking amplifiers. On the other hand, the *N*-stage Doherty amplifier generates *N* maximum efficiency points along the output power level using $(N-1)$ peaking cells with $(N-1)$ load-modulation circuits. Therefore, the $N$-stage Doherty amplifier maintains the high-efficiency characteristics over a broad power range compared with the $N$-way Doherty amplifier. The back-off levels, where the $N$-stage Doherty amplifier delivers the maximum efficiency, are adjusted by the number of stages and the size ratios of the carrier amplifier and peaking amplifiers. Since the modulation circuits are turned on sequentially, the gate-bias voltages of the later turned-on stages are very low, creating the severe current generation problem.

There are two kinds of three-stage Doherty amplifier architectures. One topology (three-stage I) consisted of a carrier amplifier based on a conventional Doherty amplifier and one more peaking amplifier. The other topology (three-stage II) has one carrier amplifier, and one peaking amplifier consisted of a conventional Doherty amplifier. For the three-stage I Doherty amplifier, initially, the carrier Doherty amplifier operates like a conventional Doherty amplifier. At a higher power, the additional peaking amplifier is turned on and modulates the load of the carrier Doherty amplifier. The other topology (three-stage II) operates similarly. At a low power, only the carrier amplifier is turned on, and at higher power, the peaking Doherty amplifier, which is a conventional Doherty by itself, is turned on similarly to the conventional Doherty amplifier. Both the three-stage and the three-way architectures use three unit cells, but the two peaking cells in the three-stage Doherty amplifier are sequentially turned on, while the peaking cells of the three-way Doherty amplifier are simultaneous turned on. Thus, three peak-efficiency points are formed for the three-stage II amplifier: at the two peaking turn-on points and at the peak power. The behaviors of the three-state Doherty amplifiers will be described in this section.

## 4.2.1 Three-Stage I Doherty Amplifier

The conventional three-stage Doherty amplifier architecture, that is widely known for a long time and we call "three-stage I", is shown in Fig. 4.14. The topology is one Doherty amplifier as a carrier amplifier with one additional peaking amplifier; initially, the carrier Doherty amplifier operates like a conventional Doherty amplifier, and at a higher power, the additional peaking amplifier turns on and modulates the load. It has two inverters for the sequential load modulation. The $N$-stage Doherty amplifier has $N-1$ peaking cells with $N-1$ inverters.

### *4.2.1.1 Fundamental Design Approach*

To evaluate the back-off levels where the three-stage I Doherty amplifier delivers the maximum efficiencies, the fundamental current profiles of the cells can be defined as shown in Fig. 4.15. It should be noticed that the current of the carrier amplifier is saturated since a proper solution could not find without the saturated operation. The maximum output power of the three-stage I Doherty amplifier is given by the total output current and DC drain bias voltage, $V_{dc}$:

$$P_{\text{Out, max}} = \frac{1}{2} \cdot V_{dc} \cdot I_{\text{Out}} \tag{4.21}$$

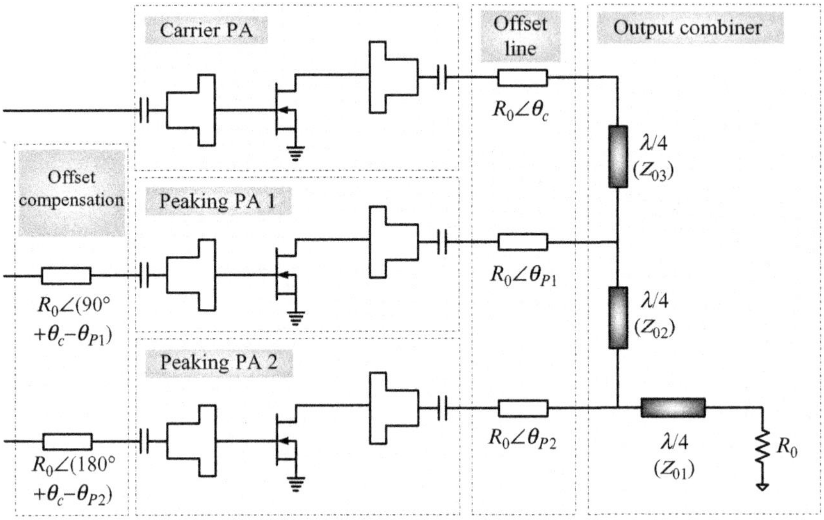


**Fig. 4.14** Topology of the three-stage I Doherty amplifier.

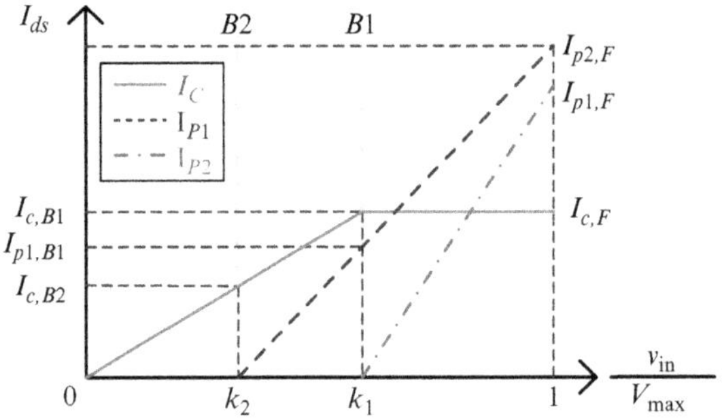


**Fig. 4.15** Fundamental current profiles of the unit cells of the three-stage I Doherty amplifier.

The maximum fundamental currents of the unit cells (carrier cell, peaking cell 1, and peaking cell 2), $I_{C,F}$, $I_{P1,F}$, and $I_{P2,F}$, follow their sizes:

$$I_{C,F} : I_{P1,F} : I_{P2,F} = 1 : m_1 : m_2 \tag{4.22}$$

$$\begin{aligned} I_{Out} &= I_{C,F} + I_{P1,F} + I_{P2,F} = I_{C,F} \cdot (1 + m_1 + m_2) \\ &= I_{P1,F} \cdot (1 + m_1 + m_2)/m_1 \\ &= I_{P2,F} \cdot (1 + m_1 + m_2)/m_2 \end{aligned} \tag{4.23}$$

From Eqs. (4.23), (4.21), the current levels are determined as

$$I_{C,F} = \frac{2 \cdot P_{Out,\,max}}{V_{dc} \cdot (1 + m_1 + m_2)} \tag{4.24}$$

$$I_{P1,F} = \frac{2 \cdot m_1 \cdot P_{\mathrm{Out,max}}}{V_{dc} \cdot (1 + m_1 + m_2)} \tag{4.25}$$

$$I_{P2,F} = \frac{2 \cdot m_2 \cdot P_{\mathrm{Out,max}}}{V_{dc} \cdot (1 + m_1 + m_2)} \tag{4.26}$$

The fundamental output current amplitudes at each back-off output power level with the maximum efficiency, as shown in Fig. 4.15, are calculated. The current levels at the first back-off point are given by

$$I_{C,B1} = I_{C,F} = \frac{2 \cdot P_{\mathrm{Out,max}}}{V_{dc} \cdot (1 + m_1 + m_2)} \tag{4.27}$$

$$I_{P1,B1} = \frac{k_1 - k_2}{1 - k_2} \cdot I_{P1,F} = \frac{k_1 - k_2}{1 - k_2} \cdot \frac{2 \cdot m_1 \cdot P_{\mathrm{Out,max}}}{V_{dc} \cdot (1 + m_1 + m_2)} \tag{4.28}$$

$$I_{P2,B1} = 0$$

where $k_1$ and $k_2$ are the turn-on voltages at the peak-efficiency points as indicated in Fig. 4.15. The current levels at the second back-off point are given by

$$I_{C,B2} = \frac{k_2}{k_1} \cdot I_{C,F} = \frac{k_2}{k_1} \cdot \frac{2 \cdot P_{\mathrm{Out,max}}}{V_{dc} \cdot (1 + m_1 + m_2)} \tag{4.29}$$

$$I_{P1,B2} = I_{P2,B2} = 0$$

Thus, the back-off output power levels are given by

$$P_{\mathrm{Out},B1} = \frac{1}{2} \cdot V_{dc} \cdot I_{C,B1} + \frac{1}{2} \cdot V_{dc} \cdot I_{P1,B1} \tag{4.30}$$

$$P_{\mathrm{Out},B2} = \frac{1}{2} \cdot V_{dc} \cdot I_{C,B2} \tag{4.31}$$

Eqs. (4.30), (4.31) can be represented in different expressions using the maximum output power in Eq. (4.21), assuming that the power is linearly increased with the input power:

$$P_{\mathrm{Out},B1} = {k_1}^2 \cdot P_{\mathrm{Out,max}} \tag{4.32}$$

$$P_{\mathrm{Out},B2} = {k_2}^2 \cdot P_{\mathrm{Out,max}} \tag{4.33}$$

Using Eqs. (4.30)–(4.33), the back-off voltage levels of the three-stage I Doherty amplifier, $k_1$ and $k_2$, are derived as

$$k_2 = \frac{1 + m_1}{1 + m_1 + m_2}, \quad k_1 = \frac{1}{1 + m_1} \tag{4.34}$$

As shown in Eq. (4.34), the back-off level of the three-stage I Doherty amplifier can be selected by changing the size ratios between the carrier cell and the peaking cells. Under the same assumption used previously that all of the cells are at class B mode after turned on and have proper transconductances, the fundamental currents of the unit cells can be calculated as functions of the input voltage using Eq. (4.34):

$$\begin{aligned} I_{1,C}(v_{in}) &= I_{C,F}\cdot\frac{v_{in}}{k_1\cdot V_{max}}, && 0<\frac{v_{in}}{V_{max}}<k_1 \\ &= I_{C,F}, && k_1<\frac{v_{in}}{V_{max}}<1 \end{aligned} \tag{4.35}$$

$$\begin{aligned} I_{1,P1}(v_{in}) &= 0, && 0<\frac{v_{in}}{V_{max}}<k_2 \\ &= \frac{I_{P1,F}}{1-k_2}\cdot\left[\frac{v_{in}}{V_{max}}-k_2\right], && k_2<\frac{v_{in}}{V_{max}}<1 \end{aligned} \tag{4.36}$$

$$\begin{aligned} I_{1,P2}(v_{in}) &= 0, && 0<\frac{v_{in}}{V_{max}}<k_1 \\ &= \frac{I_{P2,F}}{1-k_1}\cdot\left[\frac{v_{in}}{V_{max}}-k_1\right], && k_1<\frac{v_{in}}{V_{max}}<1 \end{aligned} \tag{4.37}$$

All of the unit cells are matched to the load impedance of $R_0$ when the cells deliver their full powers. At the junction, the voltage is the same for all the cells, and the impedances should be the root square of the inverse of their currents. From the property, the characteristic impedances of the quarter-wavelength transmission lines, which form the three-stage Doherty I amplifier's output combiner shown in Fig. 4.14, are determined as follows:

$$Z_{01} = R_0\cdot\sqrt{\frac{m_2}{1+m_1+m_2}} \tag{4.38}$$

$$Z_{02} = \frac{R_0}{1+m_1}\cdot\sqrt{m_1\cdot m_2} \tag{4.39}$$

$$Z_{03} = R_0\cdot\sqrt{m_1} \tag{4.40}$$

#### *4.2.1.2 Load Modulation, Efficiency and Output Power*

For the analysis of the load-modulation and efficiency characteristics, the ideal current source models of the Doherty amplifier shown in Fig. 4.16 can be used. At the region of "$0\sim k_2$," only the carrier amplifier operates, and at the region of "$k_2\sim k_1$," the carrier cell and peaking cell 1 operate. All of the cells are turned on at the region of "$k_1\sim 1$."

In Fig. 4.16A, the current source model of the three-stage I Doherty amplifier at the region of "$0\sim k_2$" is shown. The load impedance at the current source of the carrier amplifier can be derived as

$$\therefore R_{C,\sim k2}(v_{in}) = \frac{{Z_{01}}^2\cdot {Z_{03}}^2}{{Z_{02}}^2\cdot R_0} \tag{4.41}$$

The drain efficiency below the second back-off region can be calculated using the RF power and DC power that are given by

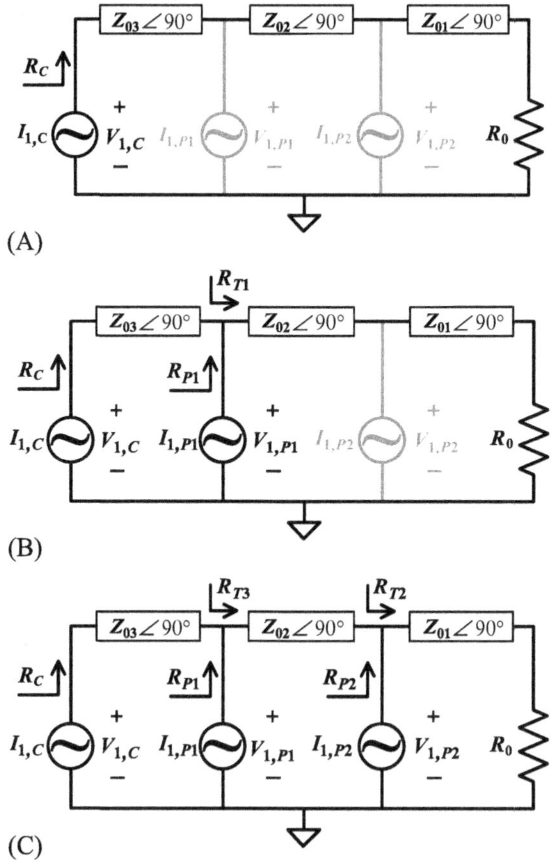


**Fig. 4.16** Ideal current source expression of the three-stage I Doherty amplifier: (A) second back-off region ($v_{in}/V_{max} = k_2$), (B) first back-off region ($v_{in}/V_{max} = k_1$), and (C) full-power ($v_{in}/V_{max} = 1$) condition (*black*, turn-on state, and *gray*, turn-off state).

$$\begin{aligned} P_{RF,\sim k2}(v_{in}) &= \frac{1}{2} \cdot I_{1,C}(v_{in})^2 \cdot R_{C,\sim k2}(v_{in}) \\ &= \frac{1}{2} \cdot I_{C,F}{}^2 \cdot \left(\frac{v_{in}}{k_1 \cdot V_{max}}\right)^2 \cdot \frac{Z_{01}{}^2 \cdot Z_{03}{}^2}{Z_{02}{}^2 \cdot R_0} = \frac{1}{2} \cdot I_{C,F} \cdot V_{dc} \cdot \left(\frac{v_{in}}{k_1 \cdot V_{max}}\right)^2 \cdot \left(\frac{Z_{01} \cdot Z_{03}}{Z_{02} \cdot R_0}\right)^2 \end{aligned} \tag{4.42}$$

$$\begin{aligned} P_{DC,\sim k2}(v_{in}) &= I_{DC,C}(v_{in}) \cdot V_{dc} = \frac{2}{\pi} \cdot I_{1,C}(v_{in}) \cdot V_{dc} \\ &= \frac{2}{\pi} \cdot I_{C,F} \cdot \frac{v_{in}}{k_1 \cdot V_{max}} \cdot V_{dc} \end{aligned} \tag{4.43}$$

$$DE_{,\sim k2}(v_{in}) = \frac{P_{RF}(v_{in})}{P_{DC}(v_{in})} = \frac{\pi}{4} \cdot \left(\frac{v_{in}}{k_1 \cdot V_{max}}\right) \cdot \left(\frac{Z_{01} \cdot Z_{03}}{Z_{02} \cdot R_0}\right)^2 \tag{4.44}$$

Fig. 4.16B represents the region of "$k_2 \sim k_1$," where carrier cell and peaking cell 1 deliver their fundamental currents to the load. The fundamental drain current ratio provided by the carrier cell and peaking cell 1 is defined as

$$\therefore \delta_1(v_{in}) = \frac{I_{1,P1}(v_{in})}{I_{1,C}(v_{in})} \tag{4.45}$$

The load impedances at the voltage nodes shown in Fig. 4.16B can be calculated using the active load-pull principle:

$$R_{T1,\sim k1} = \frac{Z_{02}^2}{Z_{01}^2} \cdot R_0 \tag{4.46}$$

$$R_{C,\sim k1}(v_{in}) = \frac{Z_{03}^2}{[1+\delta_1(v_{in})] \cdot R_{T1,\sim k1}} = \frac{Z_{01}^2 \cdot Z_{03}^2}{Z_{02}^2 \cdot R_0 \cdot [1+\delta_1(v_{in})]} \tag{4.47}$$

$$R_{P1,\sim k1}(v_{in}) = \left[1 + \frac{1}{\delta_1(v_{in})}\right] \cdot \frac{Z_{02}^2}{Z_{01}^2} \cdot R_0 \tag{4.48}$$

The drain efficiency below the first back-off region can be calculated the same way with the previous case:

$$\begin{aligned} P_{RF,\sim k1}(v_{in}) &= \frac{1}{2} \cdot I_{1,C}(v_{in})^2 \cdot R_{C,\sim k1}(v_{in}) + \frac{1}{2} \cdot I_{1,P1}(v_{in})^2 \cdot R_{P1,\sim k1}(v_{in}) \\ &= \frac{1}{2} \cdot I_{C,F} \cdot V_{dc} \left\{ \left(\frac{v_{in}}{k_1 \cdot V_{max}}\right)^2 \cdot \frac{Z_{01}^2 \cdot Z_{03}^2}{Z_{02}^2 \cdot R_0^2 \cdot [1+\delta_1(v_{in})]} + \left(\frac{1}{1-k_2}\right)^2 \cdot m_1 \cdot \right. \\ &\quad \left. \left(\frac{v_{in}}{V_{max}} - k_2\right)^2 \cdot \left[1 + \frac{1}{\delta_1(v_{in})}\right] \cdot \frac{Z_{02}^2}{Z_{01}^2} \right\} \end{aligned} \tag{4.49}$$

$$\begin{aligned} P_{DC,\sim k1}(v_{in}) &= \frac{2}{\pi} \cdot [I_{1,C}(v_{in}) + I_{1,P1}(v_{in})] \cdot V_{dc} \\ &= \frac{2}{\pi} \cdot I_{C,F} \cdot V_{dc} \cdot \left[\left(\frac{1}{k_1} + \frac{m_1}{1-k_2}\right) \cdot \frac{v_{in}}{V_{max}} - \frac{k_2 \cdot m_1}{1-k_2}\right] \end{aligned} \tag{4.50}$$

$$DE_{,\sim k1}(v_{in}) = \frac{\pi}{4} \cdot \frac{\left(\frac{v_{in}}{k_1 \cdot V_{max}}\right)^2 \cdot \frac{Z_{01}^2 \cdot Z_{03}^2}{Z_{02}^2 \cdot R_0^2 \cdot [1+\delta_1(v_{in})]} + \left(\frac{1}{1-k_2}\right)^2 \cdot m_1 \cdot \left(\frac{v_{in}}{V_{max}} - k_2\right)^2 \cdot \left[1 + \frac{1}{\delta_1(v_{in})}\right] \cdot \frac{Z_{02}^2}{Z_{01}^2}}{\left[\left(\frac{1}{k_1} + \frac{m_1}{1-k_2}\right) \cdot \frac{v_{in}}{V_{max}} - \frac{k_2 \cdot m_1}{1-k_2}\right]} \tag{4.51}$$

Fig. 4.16C represents the region of "$k_1 \sim 1$," where the current sources of all the cells supply their fundamental currents to the load. The load impedances at the each voltage node can be also calculated in the same way as before:

$$R_{T2,\sim1} = \frac{Z_{01}^2}{R_o} \tag{4.52}$$

$$R_{P2,\sim1}(v_{\text{in}}) = \left[1 + \frac{1}{\delta_2(v_{\text{in}})}\right] \cdot R_{T2,\sim1} = \left[1 + \frac{1}{\delta_2(v_{\text{in}})}\right] \cdot \frac{Z_{01}^2}{R_0} \tag{4.53}$$

$$R_{T3,\sim1}(v_{\text{in}}) = \frac{Z_{02}^2}{[1+\delta_2(v_{\text{in}})] \cdot R_{T2,\sim1}} = \frac{Z_{02}^2 \cdot R_0}{Z_{01}^2 \cdot [1+\delta_2(v_{\text{in}})]} \tag{4.54}$$

$$R_{C,\sim1}(v_{\text{in}}) = \frac{Z_{03}^2}{[1+\delta_1(v_{\text{in}})] \cdot R_{T3,\sim1}} = \frac{Z_{01}^2 \cdot Z_{03}^2}{Z_{02}^2 \cdot R_0} \cdot \frac{1+\delta_2(v_{\text{in}})}{1+\delta_1(v_{\text{in}})} \tag{4.55}$$

$$R_{P1,\sim1}(v_{\text{in}}) = \left(1 + \frac{1}{\delta_1(v_{\text{in}})}\right) \cdot R_{T3,\sim1} = \frac{Z_{02}^2 \cdot R_0}{Z_{01}^2} \cdot \frac{1+\delta_1(v_{\text{in}})}{\delta_1(v_{\text{in}}) \cdot [1+\delta_2(v_{\text{in}})]} \tag{4.56}$$

where $\delta_2(v_{\text{in}})$ is the fundamental drain current ratio between the currents provided by the carrier cell and peaking cell 1 and by the peaking cell 2, which is given by

$$\delta_2(v_{\text{in}}) = \frac{I_{1,P2}(v_{\text{in}})}{I_{1,C}(v_{\text{in}}) + I_{1,P1}(v_{\text{in}})} \tag{4.57}$$

The drain efficiency up to the full-power state can be calculated using the RF power and DC power:

$$\begin{aligned}
&\therefore P_{RF,\sim1}(v_{\text{in}}) \\
&= \frac{1}{2} \cdot I_{C,F} \cdot V_{dc} \cdot \\
&\left\{ \begin{array}{c} \left(\frac{Z_{01} \cdot Z_{03}}{Z_{02} \cdot R_0}\right)^2 \cdot \frac{1+\delta_2(v_{\text{in}})}{1+\delta_1(v_{\text{in}})} + \left(\frac{1}{1-k_2}\right)^2 \cdot m_1 \cdot \left(\frac{v_{\text{in}}}{V_{\max}} - k_2\right)^2 \cdot \frac{Z_{02}^2}{Z_{01}^2} \cdot \frac{1+\delta_1(v_{\text{in}})}{\delta_1(v_{\text{in}}) \cdot [1+\delta_2(v_{\text{in}})]} \\ + \left(\frac{1}{1-k_1}\right)^2 \cdot m_2 \cdot \left(\frac{v_{\text{in}}}{V_{\max}} - k_1\right)^2 \cdot \left[1 + \frac{1}{\delta_2(v_{\text{in}})}\right] \cdot \frac{Z_{01}^2}{R_0^2} \end{array} \right\}
\end{aligned} \tag{4.58}$$

$$P_{DC,\sim1}(v_{\text{in}}) = \frac{2}{\pi} \cdot I_{C,F} \cdot V_{dc} \cdot \left[1 + \frac{m_1}{1-k_2} \cdot \left(\frac{v_{\text{in}}}{V_{\max}} - k_2\right) + \frac{m_2}{1-k_1} \cdot \left(\frac{v_{\text{in}}}{V_{\max}} - k_1\right)\right] \tag{4.59}$$

$$\begin{aligned}
&DE_{,\sim1}(v_{\text{in}}) = \frac{\pi}{4} \cdot \\
&\frac{\left\{ \begin{array}{c} \left(\frac{Z_{01} \cdot Z_{03}}{Z_{02} \cdot R_0}\right)^2 \cdot \frac{1+\delta_2(v_{\text{in}})}{1+\delta_1(v_{\text{in}})} + \left(\frac{1}{1-k_2}\right)^2 \cdot m_1 \cdot \left(\frac{v_{\text{in}}}{V_{\max}} - k_2\right)^2 \cdot \frac{Z_{02}^2}{Z_{01}^2} \cdot \frac{1+\delta_1(v_{\text{in}})}{\delta_1(v_{\text{in}}) \cdot [1+\delta_2(v_{\text{in}})]} \\ + \left(\frac{1}{1-k_1}\right)^2 \cdot m_2 \cdot \left(\frac{v_{\text{in}}}{V_{\max}} - k_1\right)^2 \cdot \left[1 + \frac{1}{\delta_2(v_{\text{in}})}\right] \cdot \frac{Z_{01}^2}{R_0^2} \end{array} \right\}}{\left[1 + \frac{m_1}{1-k_2} \cdot \left(\frac{v_{\text{in}}}{V_{\max}} - k_2\right) + \frac{m_2}{1-k_1} \cdot \left(\frac{v_{\text{in}}}{V_{\max}} - k_1\right)\right]}
\end{aligned} \tag{4.60}$$

From Eqs. (4.44), (4.51), (4.60), the peak efficiency of 78.5% can be obtained (the maximum efficiency of class B mode amplifier) at the $v_{in}/V_{max}$ of $k_2$, $k_1$, and 1.

#### 4.2.1.3 Ideal Operational Behavior

Fig. 4.17 shows the ideal operational behavior of the three-stage I Doherty amplifier with "1:2:2" size ratio, where $k_2$ and $k_1$ are 0.33 and 0.6, respectively. It should be noticed that the size of the carrier amplifier is half of the peaking cells since the required current from the

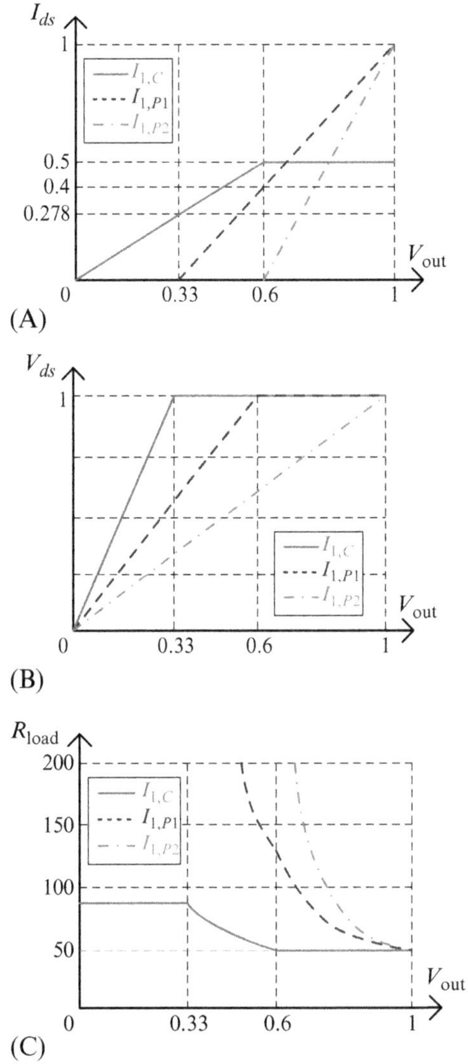


**Fig. 4.17** Ideal operational behavior of the three-stage I Doherty amplifier with "1:2:2" size ratio: (A) fundamental drain currents, (B) drain voltages, and (C) load impedances of each amplifier.

carrier amplifier is lower than those from the peaking cells. As shown in Fig. 4.17A and B, the carrier amplifier delivers the maximum efficiency for $k$ ($= V_{in}/V_{max}$) value larger than 0.33 and generates the constant output current for $k$ larger than 0.6 with a constant load impedance. It means that the carrier amplifier operates in a heavily saturated mode for $k$ larger than 0.6. Peaking cell 1 delivers the maximum efficiency for the $k$ larger than 0.6, and peaking cell 2 reaches the maximum efficiency at the peak output power. Fig. 4.17C shows the load-modulation profile; the carrier amplifier's load-modulation ratio is "$1.8R_0$:$R_0$:$R_0$," and peaking cell 1 and 2 load-modulation ratios are "open:$2.5R_0$:$R_0$" and "open:open:$R_0$," respectively. Fig. 4.18 illustrates the dynamic load lines of the unit

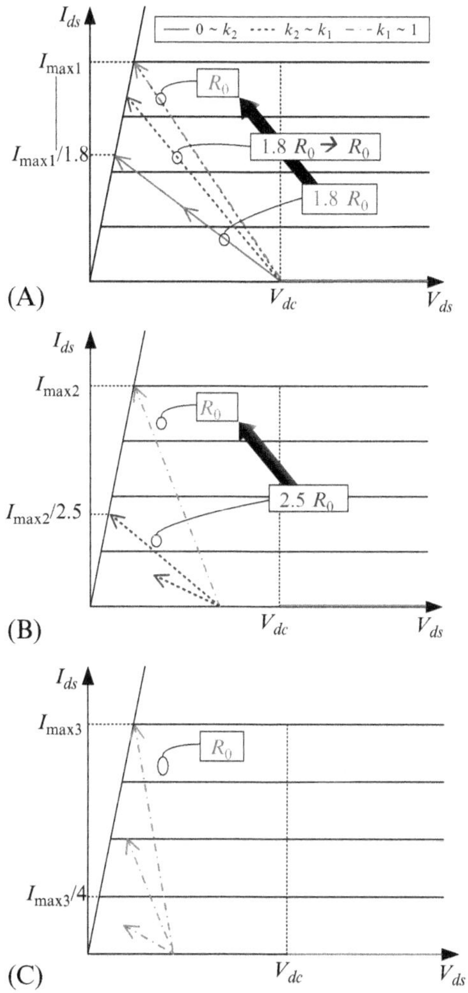


**Fig. 4.18** Load lines of the unit cells according to the output power level: (A) carrier amplifier, (B) peaking cell 1, and (C) peaking cell 2.

cells. It should be noticed that the carrier amplifier is saturated at the high-power region with the constant current and voltage. Without this saturated-mode operation, there is not any known solution for this Doherty amplifier circuit topology.

The efficiencies of the several three-stage I Doherty amplifiers are evaluated. In the design, the carrier cell is smaller than the peaking cells due to the lower current required as shown in Fig. 4.17A. Fig. 4.19 shows the efficiency profiles versus the output power. The efficiency profiles of the three-stage I Doherty amplifiers are better suited for amplification of a highly modulated signal than that of the three-way Doherty amplifier that has a large efficiency drop between the two peak-efficiency points. Table 4.2 summarizes DE of the Doherty amplifiers for amplification of the modulated WiMAX signal with 8.5 dB PAPR. The three-stage I Doherty amplifiers deliver about 10% higher efficiency compared with the three-way Doherty amplifier for amplification of the 8.5 dB PAPR signal since the amplifiers have large dynamic ranges for high efficiency operation.

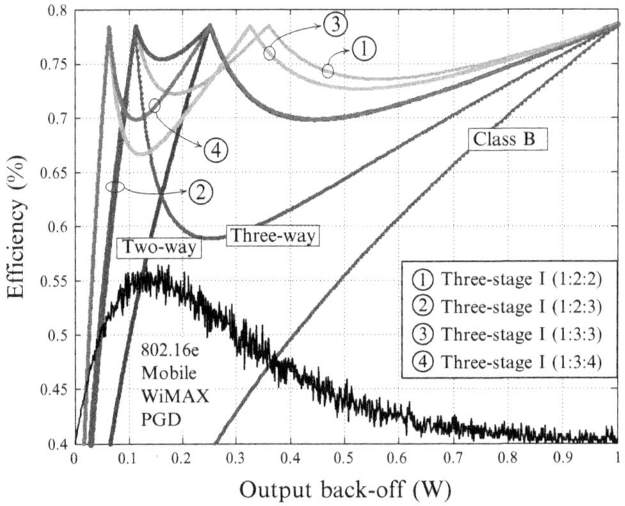


**Fig. 4.19** Simulated efficiency curves of the various three-stage I Doherty amplifiers versus the output power level.

**Table 4.2** Calculated Efficiencies of Doherty Amplifiers for WiMAX Signal With 8.5 dB PAPR

| ***N*-Way Doherty** | **Back-Off (dB)** | **Efficiency$_{avg}$ (%)** |
|---|---|---|
| Two-way | −6 | 59.02 |
| Three-way | −9.5 | 61.19 |

| **Three-Stage Cell Size Ratio** | **Back-Off at Max. Efficiency (dB)** | **Efficiency$_{avg}$ (%)** |
|---|---|---|
| 1:2:2 | −4.44/−9.5 | 69.81 |
| 1:2:3 | −6/−9.5 | 69.41 |
| 1:3:3 | −4.87/−12 | 70.46 |
| 1:3:4 | −6/−12 | 71 |

### 4.2.2 Three-Stage II Doherty Amplifier

The three-stage II Doherty amplifier is a parallel combination of one carrier amplifier and a peaking amplifier formed by one Doherty amplifier as shown in Fig. 4.20. It should be noticed that one more inverted, Z02, is added for the peaking Doherty amplifier. The fundamental current profiles of the unit cells of the three-stage II Doherty amplifier are shown in Fig. 4.21. Compared with the three-stage I Doherty amplifier, the fundamental current of the carrier amplifier does not fall into a hard saturated state, which is an indication of a linear operation.

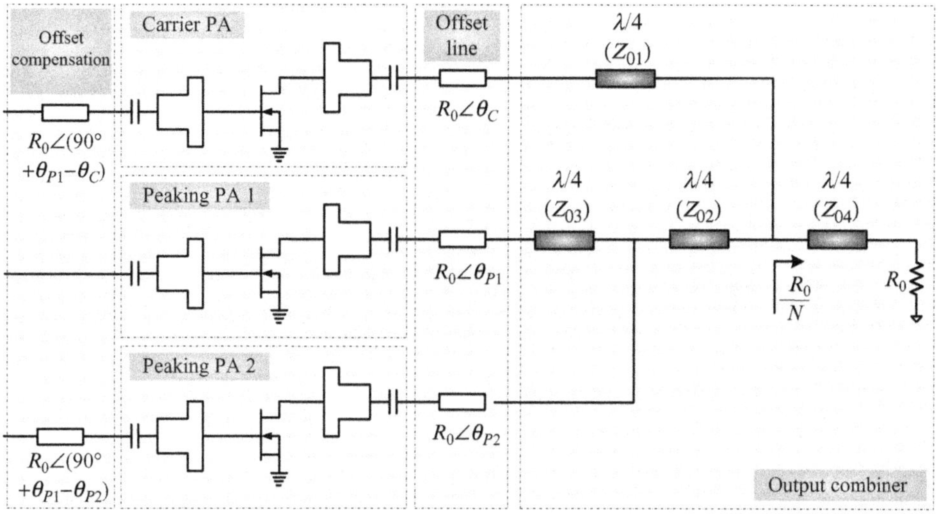


**Fig. 4.20** Topology of the three-stage II Doherty amplifier.

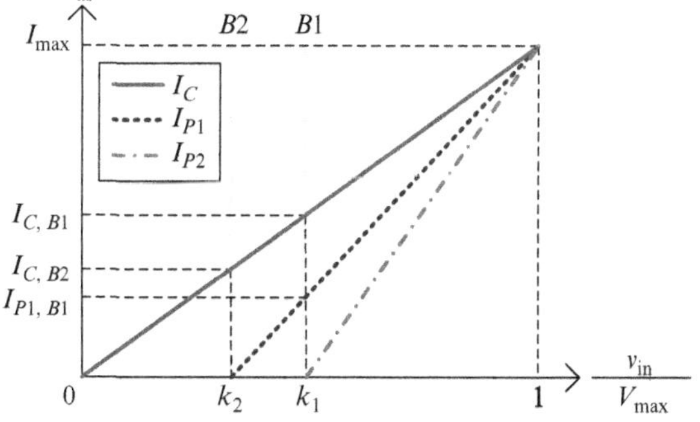


**Fig. 4.21** Current profiles of the unit cells in the three-stage II Doherty amplifier.

#### 4.2.2.1 The Peak Efficiency Points

Under the assumption that the unit cells are identical (1:1:1), the back-off levels ($k_1$ and $k_2$), where the three-stage II Doherty amplifier delivers the maximum efficiency, can be determined using the fundamental current profiles in Fig. 4.21. The maximum fundamental current of the unit cells with the same $I_{max}$ is related to the maximum output power and DC drain bias voltage, $V_{dc}$, by

$$P_{Out,max} = \frac{1}{2} \cdot V_{dc} \cdot I_{Out} = \frac{3}{2} \cdot V_{dc} \cdot I_{max} \tag{4.61}$$

The fundamental current amplitudes at the each back-off output power level defined in Fig. 4.21 are given by

$$I_{C,B1} = k_1 \cdot I_{max} = \frac{2}{3} \cdot k_1 \cdot \frac{P_{Out,max}}{V_{dc}} \tag{4.62}$$

$$I_{P1,B1} = \frac{k_1 - k_2}{1 - k_2} \cdot I_{max} = \frac{2}{3} \cdot \frac{k_1 - k_2}{1 - k_2} \cdot \frac{P_{Out,max}}{V_{dc}} \tag{4.63}$$

$$I_{P2,B1} = 0$$

$$I_{C,B2} = k_2 \cdot I_{max} = \frac{2}{3} \cdot k_2 \cdot \frac{P_{Out,max}}{V_{dc}} \tag{4.64}$$

$$I_{P1,B2} = I_{P2,B2} = 0$$

Thus, each back-off output power level can be determined as follows:

$$P_{Out,B1} = \frac{1}{2} \cdot V_{dc} \cdot I_{C,B1} + \frac{1}{2} \cdot V_{dc} \cdot I_{P1,B1} \tag{4.65}$$

$$P_{Out,B2} = \frac{1}{2} \cdot V_{dc} \cdot I_{C,B2} \tag{4.66}$$

The output powers can be represented in different expressions using the linear output power property:

$$P_{Out,B1} = k_1^2 \cdot P_{Out,max} \tag{4.67}$$

$$P_{Out,B2} = k_2^2 \cdot P_{Out,max} \tag{4.68}$$

Using Eqs. (4.65)–(4.68), the output power back-off levels of the three-stage II Doherty amplifier, $k_1$ and $k_2$, are derived:

$$k_1 = \frac{1}{2}, \quad k_2 = \frac{1}{3} \tag{4.69}$$

As shown in the Eq. (4.69), thc three-stage II Doherty amplifier delivers the maximum efficiency at the −6 dB (equivalent to 1/2) and −9.54 dB (equivalent to 1/3) back-off output power levels from the peak power and at the peak power.

#### 4.2.2.2 Load Modulation Circuit

The output-combining circuit of the three-stage II Doherty amplifier is shown in Fig. 4.22. Using the active load-pull principle, the characteristic impedances of the quarter-wave transformers in the circuit can be derived. The fundamental drain current ratios between the carrier and peaking cells are defined as

$$\delta_1(v_{\text{in}}) = \frac{I_{P1}(v_{\text{in}}) + I_{P2}(v_{\text{in}})}{I_C(v_{\text{in}})} = m_1 + m_2 = m1 \cdot [1 + \delta_2(v_{\text{in}})] \tag{4.70}$$

$$\delta_2(v_{\text{in}}) = \frac{I_{P2}(v_{\text{in}})}{I_{P1}(v_{\text{in}})} = \frac{m_2}{m_1} \tag{4.71}$$

$\delta_1$ and $\delta_2$ are calculated based on the fundamental current profiles shown in Fig. 4.21 with $k_1$ and $k_2$ of "0.5" and "0.33," respectively, and are listed in Table 4.3.

The characteristic impedances of the quarter-wave transformers, $Z_{01}$, $Z_{02}$, and $Z_{03}$, defined in Fig. 4.22, are normalized as "$MR_0$," "$QR_0$," and "$PR_0$," respectively. They are related to the load impedances of the unit cells as

$$R_C(v_{\text{in}}) = \frac{M^2 \cdot N \cdot R_0}{1 + \delta_1(v_{\text{in}})} \tag{4.72}$$

$$R_{Pt}(v_{\text{in}}) = \frac{\delta_1(v_{\text{in}})}{1 + \delta_1(v_{\text{in}})} \cdot Q^2 \cdot N \cdot R_0 \tag{4.73}$$

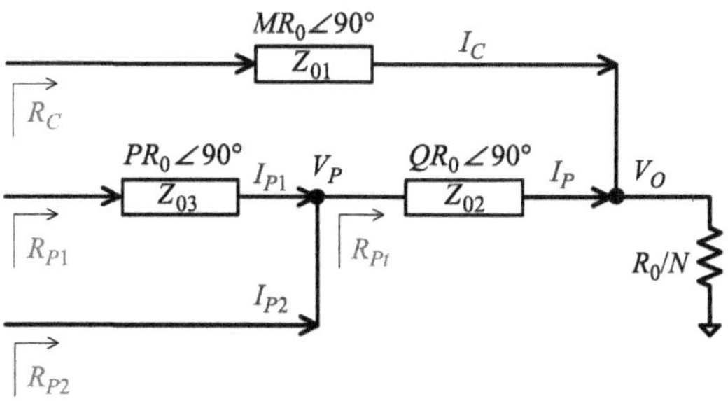


**Fig. 4.22** Output combiner of the three-stage II Doherty amplifier.

**Table 4.3** Calculated $\delta_1$ and $\delta_2$ Versus the Input Power Level

| $V_{in}/V_{max}$ | $k_2$ | $k_1$ | 1 |
|---|---|---|---|
| $I_{p1}/I_c$ | 0 | 0.5 | 1 |
| $\delta_2$ | 0 | 0 | 1 |
| $\delta_1$ | 0 | 0.5 | 2 |

$$R_{P1}(v_{\text{in}}) = \frac{[1+\delta_1(v_{\text{in}})]\cdot P^2 \cdot R_0}{\delta_1(v_{\text{in}})\cdot[1+\delta_2(v_{\text{in}})]\cdot Q^2 \cdot N} \tag{4.74}$$

$$R_{P2}(v_{\text{in}}) = \frac{\delta_1(v_{\text{in}})\cdot[1+\delta_2(v_{\text{in}})]}{\delta_2(v_{\text{in}})\cdot[1+\delta_1(v_{\text{in}})]}\cdot Q^2 \cdot N \cdot R_0 \tag{4.75}$$

Here, it is assumed that the output load of the three-stage II Doherty amplifier $R_0$ is transferred to $R_0/N$ at the combining node as shown in the figure.

Using Eqs. (4.72)–(4.75), the load impedances of the unit cells at the back-off output powers can be determined:

$$\begin{aligned} R_C(v_{\text{in}}) &= M^2 \cdot N \cdot R_0, & \frac{v_{\text{in}}}{V_{\max}} &= 0.33 \\ &= \frac{2}{3}\cdot M^2 \cdot N \cdot R_0, & \frac{v_{\text{in}}}{V_{\max}} &= 0.5 \\ &= \frac{1}{3}\cdot M^2 \cdot N \cdot R_0, & \frac{v_{in}}{V_{\max}} &= 1 \end{aligned} \tag{4.76}$$

$$\begin{aligned} R_{P1}(v_{\text{in}}) &= \infty, & \frac{v_{in}}{V_{\max}} &= 0.33 \\ &= \frac{3\cdot P^2 \cdot R_0}{Q^2 \cdot N}, & \frac{v_{\text{in}}}{V_{\max}} &= 0.5 \\ &= \frac{3\cdot P^2 \cdot R_0}{4\cdot Q^2 \cdot N}, & \frac{v_{\text{in}}}{V_{\max}} &= 1 \end{aligned} \tag{4.77}$$

$$\begin{aligned} R_{P2}(v_{\text{in}}) &= \infty, & \frac{v_{\text{in}}}{V_{\max}} &= 0.33 \\ &= \infty, & \frac{v_{\text{in}}}{V_{\max}} &= 0.5 \\ &= \frac{4}{3}\cdot Q^2 \cdot N \cdot R_0, & \frac{v_{\text{in}}}{V_{\max}} &= 1 \end{aligned} \tag{4.78}$$

Since all of the unit cells are matched to $R_0$ at $k=1$, the "$M$," "$Q$," and "$P$" are calculated as a function of the "$N$" parameter. The characteristic impedances are given by

$$Z_{01} = M \cdot R_0 = \sqrt{3/N}\cdot R_0 \tag{4.79}$$

$$Z_{02} = Q \cdot R_o = \sqrt{3/(4\cdot N)}\cdot R_0 \tag{4.80}$$

$$Z_{03} = P \cdot R_0 = R_0 \tag{4.81}$$

The load-modulation ratios of the carrier cell formulated in Eq. (4.76) are "$3R_0$":"$2R_0$":"$R_0$" and the load-modulation ratios of the peaking cell 1 and 2 are "$\infty$":"$4R_0$":"$R_0$" and "$\infty$":"$\infty$":"$R_0$," respectively.

### 4.2.2.3 Load Modulation Behavior and Efficiency

The fundamental currents of the unit cells shown in Fig. 4.21 can be derived using Eqs. (4.62)–(4.64), (4.69):

$$I_C(v_{\text{in}}) = I_{\max} \cdot \frac{v_{\text{in}}}{V_{\max}}, \quad 0 < \frac{v_{\text{in}}}{V_{\max}} < 1 \tag{4.82}$$

$$I_{P1}(v_{\text{in}}) = \begin{cases} 0, & 0 < \dfrac{v_{\text{in}}}{V_{\max}} < k_2 \\ \dfrac{I_{\max}}{1-k_2} \cdot \left[\dfrac{v_{\text{in}}}{V_{\max}} - k_2\right], & k_2 < \dfrac{v_{\text{in}}}{V_{\max}} < 1 \end{cases} \tag{4.83}$$

$$I_{P2}(v_{\text{in}}) = \begin{cases} 0, & 0 < \dfrac{v_{\text{in}}}{V_{\max}} < k_1 \\ \dfrac{I_{\max}}{1-k_1} \cdot \left[\dfrac{v_{\text{in}}}{V_{\max}} - k_1\right], & k_1 < \dfrac{v_{\text{in}}}{V_{\max}} < 1 \end{cases} \tag{4.84}$$

Using Eqs. (4.79)–(4.84), the load-modulation behavior and related efficiency characteristic of the three-stage II Doherty amplifier can be calculated. For the analysis, we have assumed that all of the unit cells are operated in class B mode. The ideal current source expressions of the Doherty amplifier for the given output power levels are illustrated in Fig. 4.23.

In the $k$ region of "0∼0.33," only the carrier amplifier is operated, and at the region of "0.33∼0.5," the carrier cell and peaking cell 1 are operated. All of the cells are turned on at the region of "0.5∼1." The ideal current source expression of the three-stage II

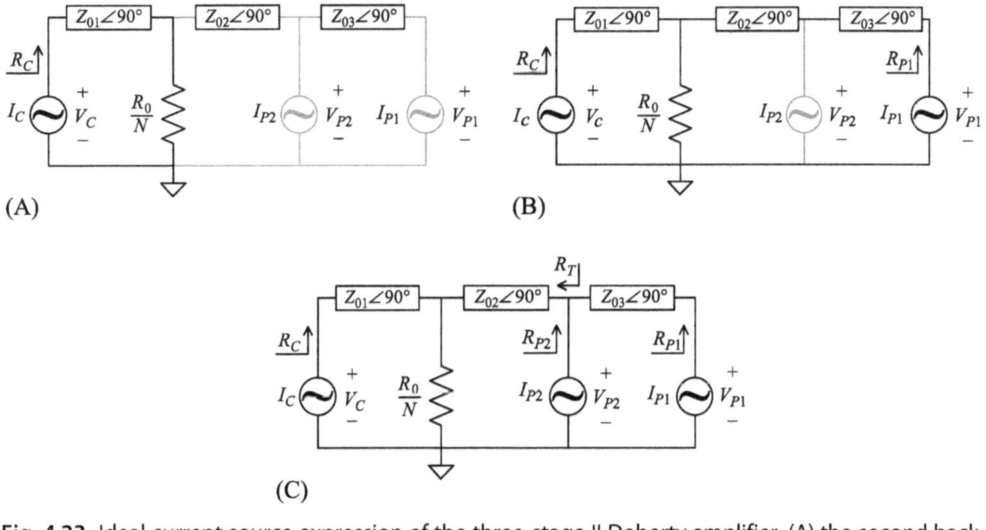

**Fig. 4.23** Ideal current source expression of the three-stage II Doherty amplifier, (A) the second back-off region, (B) first back-off region, and (C) full-power condition (*black*, turn-on state and *gray*, turn-off state).

Doherty amplifier at the region of "0 ~ 0.33" is shown in Fig. 4.23A. The load impedance at the current source of the carrier cell can be written as

$$R_{C,\sim 0.33}(v_{in}) = \frac{N \cdot {Z_{01}}^2}{R_0} \tag{4.85}$$

The drain efficiency below the second back-off region can be calculated using the RF power and DC power that are given by

$$P_{RF,\sim 0.33}(v_{in}) = \frac{1}{2} \cdot I_C(v_{in})^2 \cdot R_{C,\sim k2}(v_{in})$$
$$= \frac{1}{2} \cdot {I_{max}}^2 \cdot \left(\frac{v_{in}}{V_{max}}\right)^2 \cdot \frac{N \cdot {Z_{01}}^2}{R_0} = \frac{1}{2} \cdot I_{max} \cdot V_{DC} \cdot \left(\frac{v_{in}}{V_{max}}\right)^2 \cdot N \cdot \left(\frac{Z_{01}}{R_L}\right)^2 \tag{4.86}$$

$$P_{DC,\sim 0.33}(v_{in}) = I_{DC,C}(v_{in}) \cdot V_{dc} = \frac{2}{\pi} \cdot I_C(v_{in}) \cdot V_{dc}$$
$$= \frac{2}{\pi} \cdot I_{max} \cdot V_{DC} \cdot \frac{v_{in}}{V_{max}} \tag{4.87}$$

$$DE_{,\sim 0.33}(v_{in}) = \frac{P_{RF}(v_{in})}{P_{DC}(v_{in})} = \frac{\pi}{4} \cdot \left(\frac{v_{in}}{V_{max}}\right) \cdot N \cdot \left(\frac{Z_{01}}{R_L}\right)^2 \tag{4.88}$$

In Fig. 4.23B representing the region of "0.33 ~ 0.5," the carrier cell and peaking cell 1 supply the fundamental current to the load. The load impedances at the current sources can be calculated using the active load-pull principle:

$$R_{C,\sim 0.5}(v_{in}) = \frac{N \cdot {Z_{01}}^2}{[1 + \delta_1(v_{in})] \cdot R_0} \tag{4.89}$$

$$R_{P1,\sim 0.5}(v_{in}) = \left[1 + \frac{1}{\delta_1(v_{in})}\right] \cdot \frac{R_0}{N} \cdot \frac{{Z_{03}}^2}{{Z_{02}}^2} \tag{4.90}$$

In the same way with the previous region, the drain efficiency below the first back-off region can be calculated:

$$P_{RF,\sim 0.5}(v_{in}) = \frac{1}{2} \cdot I_C(v_{in})^2 \cdot R_{C,\sim 0.5}(v_{in}) + \frac{1}{2} \cdot I_{P1}(v_{in})^2 \cdot R_{P1,\sim 0.5}(v_{in})$$
$$= \frac{1}{2} \cdot I_{max} \cdot V_{DC} \left\{ \left(\frac{v_{in}}{V_{max}}\right)^2 \cdot \frac{N}{[1 + \delta_1(v_{in})]} \cdot \left(\frac{Z_{01}}{R_0}\right)^2 + \left(\frac{1}{1 - k_2}\right)^2 \cdot \left(\frac{v_{in}}{V_{max}} - k_2\right)^2 \cdot \right.$$
$$\left. \left[1 + \frac{1}{\delta_1(v_{in})}\right] \cdot \frac{{Z_{03}}^2}{N \cdot {Z_{02}}^2} \right\} \tag{4.91}$$

$$P_{DC,\sim 0.5}(v_{in}) = \frac{2}{\pi} \cdot [I_C(v_{in}) + I_{P1}(v_{in})] \cdot V_{dc}$$
$$= \frac{2}{\pi} \cdot I_{max} \cdot V_{DC} \cdot \left[\left(1 + \frac{1}{1 - k_2}\right) \cdot \frac{v_{in}}{V_{max}} - \frac{k_2}{1 - k_2}\right] \tag{4.92}$$

$$DE_{,\sim 0.5}(v_{in}) = \frac{\pi}{4} \cdot \frac{\left(\frac{v_{in}}{V_{max}}\right)^2 \cdot \frac{N}{[1+\delta_1(v_{in})]} \cdot \left(\frac{Z_{01}}{R_0}\right)^2 + \left(\frac{1}{1-k_2}\right)^2 \cdot \left(\frac{v_{in}}{V_{max}} - k_2\right)^2 \cdot \left[1 + \frac{1}{\delta_1(v_{in})}\right] \cdot \frac{Z_{03}{}^2}{N \cdot Z_{02}{}^2}}{\left(1 + \frac{1}{1-k_2}\right) \cdot \frac{v_{in}}{V_{max}} - \frac{k_2}{1-k_2}} \tag{4.93}$$

In Fig. 4.23C representing the region of "0.5 ~ 1," all the current sources of the unit cells supply the fundamental current to the load. The load impedances at the nodes defined in Fig. 4.23C can be calculated in the same way:

$$R_{T,\sim 1}(v_{in}) = \frac{Z_{02}{}^2}{\left[1 + \frac{1}{\delta_1(v_{in})}\right] \cdot \frac{R_0}{N}} = \frac{\delta_1(v_{in}) \cdot N \cdot Z_{02}{}^2}{[1+\delta_1(v_{in})] \cdot R_0} \tag{4.94}$$

$$R_{P2,\sim 1}(v_{in}) = \left[1 + \frac{1}{\delta_2(v_{in})}\right] \cdot R_{T,\sim 1} = \frac{[1+\delta_2(v_{in})] \cdot \delta_1(v_{in})}{\delta_2(v_{in}) \cdot [1+\delta_1(v_{in})]} \cdot \frac{N \cdot Z_{02}{}^2}{R_0} \tag{4.95}$$

$$R_{P1,\sim 1}(v_{in}) = \frac{Z_{03}{}^2}{[1+\delta_2(v_{in})] \cdot R_{T,\sim 1}} = \frac{1+\delta_1(v_{in})}{\delta_1(v_{in}) \cdot [1+\delta_2(v_{in})]} \cdot \frac{R_o \cdot Z_{03}{}^2}{N \cdot Z_{02}{}^2} \tag{4.96}$$

$$R_{C,\sim 1}(v_{in}) = \frac{Z_{01}{}^2}{[1+\delta_1(v_{in})] \cdot \frac{R_0}{N}} = \frac{N \cdot Z_{01}{}^2}{[1+\delta_1(v_{in})] \cdot R_0} \tag{4.97}$$

The drain efficiency up to the full-power state can be calculated using the RF power and DC power:

$$P_{RF,\sim 1}(v_{in}) = \frac{1}{2} \cdot I_{max} \cdot V_{DC} \cdot \left\{ \begin{array}{l} \frac{N}{[1+\delta_1(v_{in})]} \cdot \left(\frac{Z_{01}}{R_0}\right)^2 + \left(\frac{1}{1-k_2}\right)^2 \cdot \left(\frac{v_{in}}{V_{max}} - k_2\right)^2 \cdot \frac{1+\delta_1(v_{in})}{\delta_1(v_{in}) \cdot [1+\delta_2(v_{in})]} \cdot \frac{Z_{03}{}^2}{N \cdot Z_{02}{}^2} \\ + \left(\frac{1}{1-k_1}\right)^2 \cdot \left(\frac{v_{in}}{V_{max}} - k_1\right)^2 \cdot \frac{[1+\delta_2(v_{in})] \cdot \delta_1(v_{in})}{\delta_2(v_{in}) \cdot [1+\delta_1(v_{in})]} \cdot N \cdot \left(\frac{Z_{02}}{R_0}\right)^2 \end{array} \right\} \tag{4.98}$$

$$P_{DC,\sim 1}(v_{in}) = \frac{2}{\pi} \cdot I_{max} \cdot V_{DC} \cdot \left[\frac{v_{in}}{V_{max}} + \frac{1}{1-k_2} \cdot \left(\frac{v_{in}}{V_{max}} - k_2\right) + \frac{1}{1-k_1} \cdot \left(\frac{v_{in}}{V_{max}} - k_1\right)\right] \tag{4.99}$$

$$DE_{,\sim 1}(v_{in}) = \frac{\pi}{4} \cdot \frac{\left\{ \begin{array}{l} \dfrac{N}{[1+\delta_1(v_{in})]} \cdot \left(\dfrac{Z_{01}}{R_0}\right)^2 + \left(\dfrac{1}{1-k_2}\right)^2 \cdot \left(\dfrac{v_{in}}{V_{max}} - k_2\right)^2 \cdot \dfrac{1+\delta_1(v_{in})}{\delta_1(v_{in})\cdot[1+\delta_2(v_{in})]} \cdot \dfrac{{Z_{03}}^2}{N \cdot {Z_{02}}^2} \\ + \left(\dfrac{1}{1-k_1}\right)^2 \cdot \left(\dfrac{v_{in}}{V_{max}} - k_1\right)^2 \cdot \dfrac{[1+\delta_2(v_{in})]\cdot\delta_1(v_{in})}{\delta_2(v_{in})\cdot[1+\delta_1(v_{in})]} \cdot N \cdot \left(\dfrac{Z_{02}}{R_0}\right)^2 \end{array} \right\}}{\left[\dfrac{v_{in}}{V_{max}} + \dfrac{1}{1-k_2}\cdot\left(\dfrac{v_{in}}{V_{max}} - k_2\right) + \dfrac{1}{1-k_1}\cdot\left(\dfrac{v_{in}}{V_{max}} - k_1\right)\right]} \tag{4.100}$$

From Eqs. (4.88), (4.93), and (4.100), the maximum efficiencies of a class B mode amplifier, 78.5%, are obtained at $k_2$, $k_1$, and 1, respectively.

Fig. 4.24 shows the ideal load-modulation behavior of the three-stage II Doherty amplifier. As shown in Fig. 4.24A and B, the carrier amplifier operates similarly to the carrier amplifier in a standard Doherty amplifier, that is, operates as a class B amplifier at $k$ smaller than $k_2$ and delivers the maximum efficiency at $k_2$ of 0.33. The efficiency is maintained, and its output power is increased linearly in the region with $k$ larger than $k_2$ since the current is increased linearly while the voltage is maintained due to the

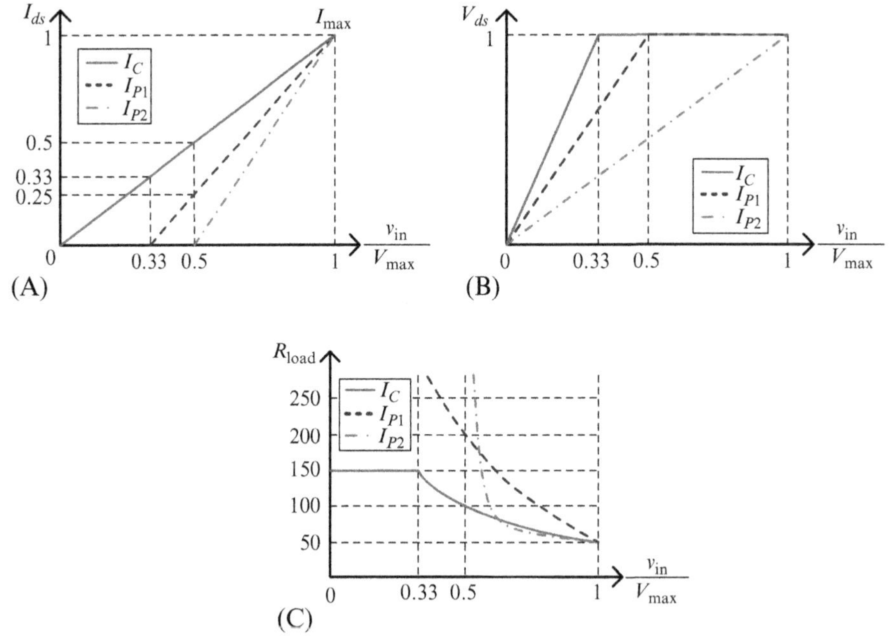


**Fig. 4.24** Ideal operation of the three-stage II Doherty amplifier with "1:1:1" size ratio: (A) fundamental drain currents, (B) drain voltages, and (C) load impedances.

modulated load impedance shown in Fig. 4.24C. This behavior is quite different from that of the three-stage I Doherty amplifier. The peaking cell 1 turns on at $k = 0.33$, operating as a class B amplifier. It delivers the maximum efficiency at $k_1$ of 0.5, with the same voltage but lower current than those of the carrier amplifier due to the larger load impedance. In the region of $k$ larger than $k_1$, the power is linearly increased with the maximum efficiency. The peaking cell 2 reaches the maximum efficiency at the peak output power, similarly to the peaking amplifier in a standard Doherty amplifier. Fig. 4.25 illustrates the load lines of the cells. It should be noticed that there is no saturated operation region.

#### *4.2.2.4 Three-Stage II Doherty Amplifier With Asymmetric Size Ratio*

In the previous section, the three-stage II Doherty amplifier with 1:1:1 size ratio is discussed. The back-off levels ($k_1$ and $k_2$) of the three-stage II Doherty amplifier can be changed by adjusting the size ratio of the unit cells, similarly to the three-stage I Doherty amplifier. For the analysis of the different-size Doherty amplifier, the fundamental current profiles as shown in Fig. 4.26 are used. The maximum fundamental currents of the cells can be determined by the maximum output power $P_{Out,\ max}$ and DC drain bias voltage, $V_{dc}$. The $P_{Out,\ max}$ is given by

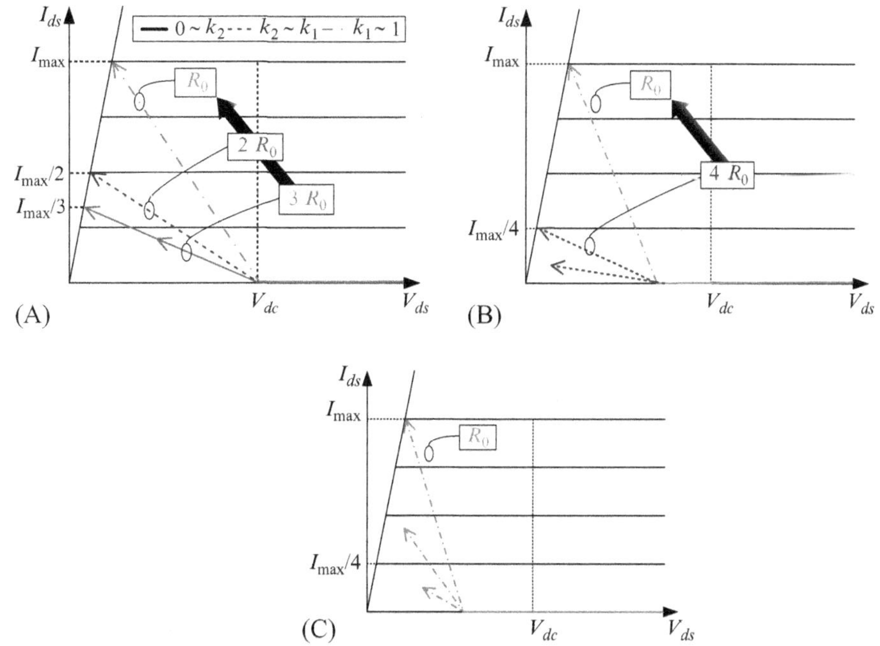


Fig. 4.25 Load lines of the unit cell amplifiers according to the output power level: (A) carrier cell, (B) peaking cell 1, and (C) peaking cell 2.

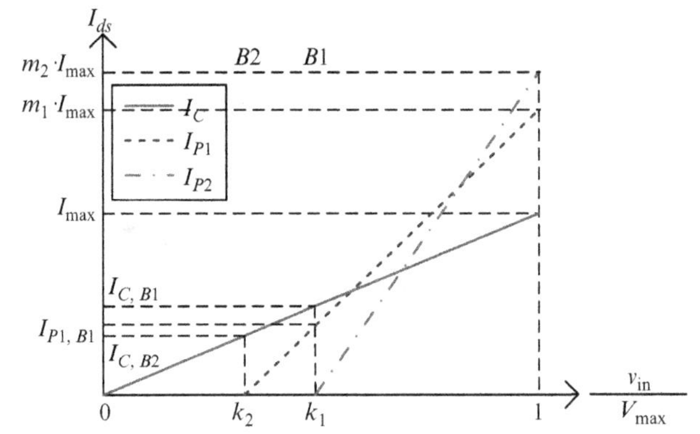


**Fig. 4.26** Fundamental current profiles of the three-stage II Doherty amplifier with different size ratios between the unit cells.

$$P_{\text{Out,max}} = \frac{1}{2} \cdot V_{dc} \cdot (1 + m_1 + m_2) \cdot I_{\max} \tag{4.101}$$

The fundamental current amplitudes at the back-off output power levels, shown in Fig. 4.26, are given by

$$I_{C,B1} = k_1 \cdot I_{\max} = \frac{2 \cdot k_1}{1 + m_1 + m_2} \cdot \frac{P_{\text{Out,max}}}{V_{dc}} \tag{4.102}$$

$$\begin{aligned} I_{P1,B1} &= \frac{k_1 - k_2}{1 - k_2} \cdot m_1 \cdot I_{\max} = \frac{2 \cdot m_1}{1 + m_1 + m_2} \cdot \frac{k_1 - k_2}{1 - k_2} \cdot \frac{P_{\text{Out,max}}}{V_{dc}} & k_2 < v_{in}/V_{\max} < 1 \\ I_{P2,B1} &= 0 & 0 < v_{in}/V_{\max} < k_2 \end{aligned} \tag{4.103}$$

$$\begin{aligned} I_{C,B2} &= k_2 \cdot I_{\max} = \frac{2 \cdot k_2}{1 + m_1 + m_2} \cdot \frac{P_{\text{Out,max}}}{V_{dc}} & k_1 < v_{in}/V_{\max} < 1 \\ I_{P1,B2} &= I_{P2,B2} = 0 & 0 < v_{in}/V_{\max} < k_1 \end{aligned} \tag{4.104}$$

The back-off output power levels are given by

$$P_{\text{Out},B1} = \frac{1}{2} \cdot V_{dc} \cdot I_{C,B1} + \frac{1}{2} \cdot V_{dc} \cdot I_{P1,B1} \tag{4.105}$$

$$P_{\text{Out},B2} = \frac{1}{2} \cdot V_{dc} \cdot I_{C,B2} \tag{4.106}$$

The output powers can be represented in different forms using the maximum output power given in Eq. (4.101):

$$P_{\text{Out},B1} = k_1^2 \cdot P_{\text{Out,max}} \tag{4.107}$$

$$P_{\text{Out},B2} = k_2^2 \cdot P_{\text{Out,max}} \tag{4.108}$$

Using Eqs. (4.105)–(4.108), the output power back-off levels of the three-stage II Doherty amplifier, $k_1$ and $k_2$, are derived:

$$k_1 = \frac{-b + \sqrt{b^2 - 4 \cdot a \cdot c}}{2 \cdot a}, \quad k_2 = \frac{1}{1 + m_1 + m_2} \tag{4.109}$$

$$A = 1 + m_1 + m_2, \qquad B = 1 + m_2$$
$$a = (A - 1) \cdot A, \quad b = -[A - 1 + (A - B) \cdot A], \quad c = A - B$$

Similarly to the three-stage I Doherty amplifier, the back-off levels are changed with the size ratios of $m_1$ and $m_2$. To figure out the load-modulation characteristic of the three-stage II Doherty amplifier using the unit cells with different sizes, the fundamental current ratios ($\delta_1$ and $\delta_2$) between the carrier and peaking cells in Eqs. (4.70), (4.71) are redefined in Eqs. (4.102)–(4.104):

$$\begin{aligned}
\delta_1(v_{\text{in}}) &= \frac{I_{P1}(v_{\text{in}}) + I_{P2}(v_{\text{in}})}{I_C(v_{\text{in}})} \\
&= 0, && v_{\text{in}}/V_{\text{max}} = k_2 \\
&= \frac{k_1 - k_2}{k_1 \cdot (1 - k_2)} \cdot m_1, && v_{\text{in}}/V_{\text{max}} = k_1 \\
&= m_1 + m_2, && v_{\text{in}}/V_{\text{max}} = 1
\end{aligned} \tag{4.110}$$

$$\begin{aligned}
\delta_2(v_{\text{in}}) &= \frac{I_{P2}(v_{\text{in}})}{I_{P1}(v_{\text{in}})} \\
&= 0, && v_{\text{in}}/V_{\text{max}} = k_2 \\
&= 0, && v_{\text{in}}/V_{\text{max}} = k_1 \\
&= \frac{m_2}{m_1}, && v_{\text{in}}/V_{\text{max}} = 1
\end{aligned} \tag{4.111}$$

Accordingly, the load impedances, defined in Fig. 4.22, of the amplifiers at the back-off output power can be determined:

$$\begin{aligned}
R_c(v_{\text{in}}) &= M^2 \cdot N \cdot R_0, && \frac{v_{\text{in}}}{V_{\text{max}}} = 0.33 \\
&= \frac{1}{1 + \left[\frac{k_1 - k_2}{k_1 \cdot (1 - k_2)}\right] \cdot m_1} \cdot M^2 \cdot N \cdot R_0, && \frac{v_{\text{in}}}{V_{\text{max}}} = 0.5 \\
&= \frac{1}{1 + m_1 + m_2} \cdot M^2 \cdot N \cdot R_0, && \frac{v_{\text{in}}}{V_{\text{max}}} = 1
\end{aligned} \tag{4.112}$$

$$\begin{aligned}
R_{P1}(v_{\text{in}}) &= \infty, && \frac{v_{\text{in}}}{V_{\text{max}}} = 0.33 \\
&= \left(1 + \frac{k_1 \cdot (1 - k_2)}{m_1 \cdot (k_1 - k_2)}\right) \cdot \frac{P^2 \cdot R_0}{Q^2 \cdot N}, && \frac{v_{\text{in}}}{V_{\text{max}}} = 0.5 \\
&= \frac{m_1 \cdot (1 + m_1 + m_2)}{(m_1 + m_2)^2} \cdot \frac{P^2 \cdot R_0}{Q^2 \cdot N}, && \frac{v_{\text{in}}}{V_{\text{max}}} = 1
\end{aligned} \tag{4.113}$$

$$R_{P2}(v_{in}) = \begin{cases} \infty, & \dfrac{v_{in}}{V_{max}} = 0.33 \\ \infty, & \dfrac{v_{in}}{V_{max}} = 0.5 \\ \dfrac{(m_1+m_2)^2}{m_2 \cdot (1+m_1+m_2)} \cdot Q^2 \cdot N \cdot R_0, & \dfrac{v_{in}}{V_{max}} = 1 \end{cases} \quad (4.114)$$

Since all of the unit amplifiers are matched to "$R_0$" at $k=1$, the "$M$," "$Q$," and "$P$" are calculated from Eqs. (4.112) to (4.114) as a function of the "$N$" parameter. Accordingly, the characteristic impedances of the transformers can be calculated using Eqs. (4.79)–(4.81):

$$Z_{01} = M \cdot R_0 = \sqrt{\frac{(1+m_1+m_2)}{N}} \cdot R_0 \quad (4.115)$$

$$Z_{02} = Q \cdot R_0 = \frac{1}{m_1+m_2} \cdot \sqrt{\frac{m_2 \cdot (1+m_1+m_2)}{N}} \cdot R_0 \quad (4.116)$$

$$Z_{03} = P \cdot R_0 = \frac{m_2}{m_1} R_0 \quad (4.117)$$

Using the above equations, the output combiner of the three-stage II Doherty amplifier can be designed with various size ratios.

#### *4.2.2.5 Calculated Efficiency Profile of the Three-Stage II Doherty Amplifier*

The efficiency curves of the three-stage II Doherty amplifiers are calculated, and the results are shown in Fig. 4.27. The efficiency curves are compared with the three-way Doherty amplifier and the three-stage I Doherty amplifier. It can be seen that the efficiency profiles are identical to the three-stage I Doherty amplifier with different size ratios. But the efficiency profile is better suited for amplification of a highly modulated signal than that of the three-way Doherty amplifier since they do not have large efficiency drop between the two peak-efficiency points. Due to the same efficiency profiles of the three-stage I amplifier (1:2:3): three-stage II amplifier (1:1:1) and three-stage I amplifier (1:3:4): three-stage II amplifier (2:3:3), their efficiencies for amplification of the modulated signal should be identical also, and they are listed in Table 4.2.

#### *4.2.2.6 Gain of the Three-Stage II Doherty Amplifier*

In Section 4.2.2, it is shown that the three-stage I Doherty amplifier cannot maintain the linear gain response with the output power since the carrier amplifier is saturated as the peaking cell 2 is turned on. However, the three-stage II Doherty amplifier provides a constant linear gain without having the saturated operation region. The linear gain means the gain of the three-stage Doherty amplifier at "$0 \sim k_2$" region, where only the carrier amplifier is operated. The linear gain is the achievable gain of the Doherty amplifier

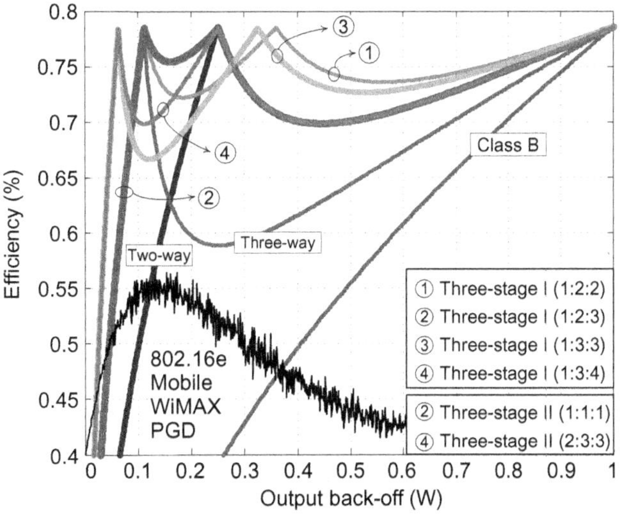


**Fig. 4.27** Simulated efficiency curves of the various three-stage II Doherty amplifiers versus the output power level. The efficiency curves of the two- and three-way and three-stage I Doherty amplifiers are also shown for comparison.

without any saturated operation, and the gain is the important target value of a linear Doherty amplifier design. To calculate the linear gain, the input power portion delivered to the carrier amplifier and the modulated load impedance should be considered.

Under the condition that all of the unit cells with different sizes have the same gain, the input power dividing ratio to the cells should follow the device size ratio ($1{:}m_1{:}m_2$) because the full powers should be generated from the unit cells at the peak-power operation. The load-modulation ratio of the carrier amplifier from $k$ of 1 to $k_2$ can be obtained from Eq. (4.112):

$$\Gamma_m = \frac{R_{catk2}}{R_{cat1}} = 1 + m_1 + m_2 \tag{4.118}$$

As the load impedance increases through the load modulation, the output power of the carrier amplifier decreases linearly since the voltage is maintained and the current is reduced by the same factor. Therefore, the power is reduced by $10{\cdot}\log(1/\Gamma_m)$ in decibel scale. However, the input power delivered to the carrier amplifier at $k_2$ is reduced by $10{\cdot}\log(1/\Gamma_m)^2$ dB from the input at the peak-power operation. Accordingly, the gain of the carrier amplifier itself at the low-power region is expanded by $G_E$, due to the larger load, which is given by

$$G_E = 10 \cdot \log(\Gamma_m) = 10 \cdot \log(1 + m_1 + m_2) \ \text{(dB)} \tag{4.119}$$

On the other hand, only a portion of the total input power to the Doherty amplifier is delivered to the carrier amplifier through the input-dividing circuit, following the size ratio:

$$D_L = 10 \cdot \log\left(\frac{1}{1 + m_1 + m_2}\right) \text{ (dB)} \tag{4.120}$$

From Eqs. (4.119), (4.120), we can see that the linear gain of the three-stage II Doherty amplifier at the region of "$0 \sim k_2$" is equal to the gain of the carrier amplifier at the peak-power operation since the higher gain due to the larger load is compensated by the input power loss that is delivered to the turn-off peaking cells:

$$\Delta \text{Gain}_L = G_E + D_L = 10 \cdot \log\left(\frac{1 + m_1 + m_2}{1 + m_1 + m_2}\right) = 0 \text{ (dB)} \tag{4.121}$$

At the peak-power operation, all the input power is amplified by the unit cells with the identical gain. It means that the three-stage II Doherty amplifier delivers the same gain at "$0 \sim k_2$" region and the peak-power operation. The gain is maintained in the other operation regions, where the $G_E$ and $D_E$ are canceled, similarly to Eq. (4.121).

### 4.2.3 Problems in Implementation of the Three-Stage Doherty Amplifiers

As we have described so far, the three-stage Doherty amplifiers can deliver very good efficiency for amplification of a highly modulated signal. But the amplifiers are not very popular due to their complicated circuit structure, and further research effort is needed for fully utilizing their capability.

The behavior of the Doherty amplifiers we have described so far assumes that all of the unit cells reach their maximum current levels at the peak output power even though their biases are different. Because the peaking cells are biased at lower voltages, they are turned on at higher input powers than the carrier cell, and their current levels are lower. Also, for the same peak current level, the fundamental component content of the lower-biased operation is smaller. Due to the smaller fundamental current levels of the peaking cells, the modulated load impedances are higher than those of the ideal operation, disturbing the proper load modulation.

Therefore, the three-stage Doherty amplifier should be selected properly. For example, the three stage II with 1:1:1 ratio has the peaking cell II turn-on voltage identical to the conventional Doherty amplifier, relaxing the problem. The peaking cells of the N-way Doherty amplifier are biased at a class C mode, but the bias increases with N. Therefore, the current problem is relaxed with N and it can be realized easily.

To get the proper load modulation from the complicated modulation circuit, an accurate uneven drive is needed, but it is very difficult to be realized. Moreover, the gain degradation can be significant due to the large portion of the input power delivered to the off-state peaking cells.

Another method is to generate the three different input signals, at the digital domain, appropriate for the three unit cells of the three-stage Doherty amplifier. The signals are delivered to the cells separately, while the combined output recovers the amplified original signal. In this case, we need to generate the new input signal with three input drivers, which is an expensive process. The signal should be synchronized properly also, making the circuit very complex.

The alternative is the gate-bias control of the peaking amplifiers described in Section 2.1.2. The gate bias of the N-way Doherty amplifier can be adapted easily to compensate the low currents of the peaking cells. But the three-stage Doherty amplifier cannot be adapted properly due to the large difference of the bias voltages. It is more difficult for the three-stage I Doherty amplifier with the saturated operation region. This process also degrades the previous power gain.

The three-stage I Doherty amplifier has Schottky turn-on problem for the GaN HEMT device. The carrier amplifier operates in the saturated mode. The load is not fully modulated to the proper low value at the high-power region. This saturated operation of the carrier amplifier causes Schottky turn-on problem for the GaN HEMT device because the carrier is driven over the full current region due to the high-input drive if the input power is not properly reduced. However, the three-stage II Doherty amplifier does not have the problem.

## FURTHER READING


[1] F.H. Raab, Efficiency of Doherty RF power amplifier system, IEEE Trans. Broadcast. BC-33 (3) (1987) 77–83.

[2] Y. Yang, et al., A fully matched N-way Doherty amplifier with optimized linearity, IEEE Trans. Microwave Theory Tech. 51 (3) (2003) 986–993.

[3] M.J. Pelk, et al., A high-efficiency 100-W GaN three-way doherty amplifier for base-station applications, IEEE Trans. Microwave Theory Tech. 56 (7) (2008) 1582–1591.

[4] D. Gustafsson, et al., A modified Doherty power amplifier with extended bandwidth and reconfigurable efficiency, IEEE Trans. Microwave Theory Tech. 61 (1) (2013) 533–542.

[5] I. Kim, et al., Optimized design of highly efficient 3-stage Doherty PA using gate adaptation, IEEE Trans. Microwave Theory Tech. 58 (10) (2010) 2562–2574.

[6] B. Kim, et al., Advanced Doherty architecture, IEEE Microw. Mag. 11 (5) (2010) 72–86.

[7] M. Iwamoto, et al., An extended Doherty amplifier with high efficiency over a wide power range, IEEE Trans. Microwave Theory Tech. 49 (12) (2001) 2472–2479.

[8] N. Srirattana, et al., Analysis and design of a high-efficiency multistage Doherty power amplifier for wireless communications, IEEE Trans. Microwave Theory Tech. 53 (3) (2005) 852–860.

CHAPTER FIVE

# Linear Doherty Power Amplifier for Handset Application

Doherty power amplifier is a good solution for amplification of a high PAPR signal as clearly seen from the popularity in the base-station amplification. But the amplifier is less popular for handset application because of the narrow bandwidth and complex circuit topology. However, those problems can be solved using elaborated circuit design works. This chapter is devoted to the Doherty amplifier for handset application. The design technique using GaAs HBT will be described in detail, and CMOS Doherty power amplifier will also be discussed.

## 5.1 INTRODUCTION

For the mobile handset application, the Doherty structure should be integrated into a single chip with a small size. The bulky input power divider and quarter-wavelength transmission line must be replaced by circuits with small sizes. Above all, the linearity of the handset Doherty power amplifier should be considered more carefully than that of the base-station amplifier because the powerful digital predistortion technique is not usually employed.

The ideal Doherty amplifier is linear because the uneven powers from the carrier- and peaking amplifiers are combined perfectly, providing the linear AM-AM characteristic as discussed in Section 1.2.2. But the real transistor responses are different from the ideal class B and C bias operations, and the nonlinear responses should be corrected to get the linear response. The class B carrier amplifier should be replaced by a linear class AB amplifier because only the carrier amplifier is operated at the low-power region. The nonlinear operation of the class C peaking amplifier should be linearized. The natural choice for linearization of the Doherty power amplifier is to compensate the gain compression of the class AB carrier amplifier and the gain expansion of the class C peaking amplifier. For the purpose, the peaking amplifier should be turned on early, and the third-order intermodulations (IM3) of the two amplifiers should be adjusted properly for the IM3 cancellation. The even harmonics should be suppressed by the harmonic short load since the harmonics can generates the noncancelable IM3 component through the interaction with the fundamental component. Also, the uneven drive is necessary to get the ideal Doherty amplifier operation.

In this chapter, detailed design processes of a linear Doherty power amplifier for handset will be introduced. For a compact design, the quarter-wavelength inverter is

*Doherty Power Amplifiers*
https://doi.org/10.1016/B978-0-12-809867-7.00005-3

replaced by a proper lumped quarter-wave transformer, and a bulky input power divider is replaced by a direct input-power-dividing circuit. The direct dividing circuit can optimize the load modulation characteristic through the proper uneven input drive and reduces the number of the circuit components. The gain modulation properties of the carrier and peaking amplifiers are analyzed, and a method to get a linear AM-AM characteristic is introduced. A proper lumped quarter-wave transformer and harmonic terminations are introduced to improve the linearity. Based on the concept, a linear Doherty power amplifier is designed using heterojunction bipolar transistor (HBT), since HBT is the most favored technology for handset power amplifiers.

The same technique can be applied to a CMOS Doherty power amplifier. In the CMOS power amplifier, the output power is voltage-combined using a transformer. The operation behavior of the CMOS Doherty amplifier based on the transformer is described in Section 1.4.1.

# 5.2 DESIGN OF LINEAR DOHERTY POWER AMPLIFIER

## 5.2.1 Load Modulation of Doherty Amplifier Based on HBT

In the Doherty amplifier at a low power operation, the input power is divided by a 3 dB coupler, and the 3 dB lower input power is delivered to the carrier, producing a 3 dB loss. Therefore, the overall gain of Doherty amplifier is equivalent to the gain with $R_{OPT}$ load. At the peak power operation, the carrier and peaking amplifiers become a two-way current power combining structure with $R_{OPT}$ load, and the gain is the same as at a low-power level. In between the two power levels, due to the nice asymmetrical power combining, the gain is maintained as we have described in Section 1.2.2. Due to the constant gain characteristic, the Doherty amplifier is a linear amplifier. However, in the practical design of the Doherty power amplifier using an HBT, the ideal gain characteristic is not produced.

### *5.2.1.1 Gain Modulation of Carrier Amplifier*

Fig. 5.1 shows the nonlinear equivalent circuit model of an HBT. In the HBT model, $C_{be}$, $g_{be}$, and $g_m$ are strong nonlinear components, varying as an exponential function of $V_{be}$. But their responses are canceled out since the output current $g_m V_{be}$ is proportional to $g_m/(j\omega C_{be}+g_{be})$. And the main nonlinear component of an HBT is $C_{bc}$.

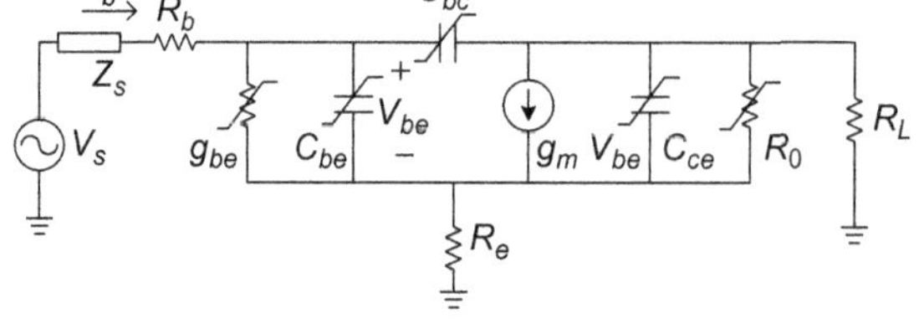


**Fig. 5.1** Nonlinear equivalent circuit model of an HBT.

$C_{bc}$ is the base-collector depletion layer capacitance, and it is decreased with $V_{bc}$ because of the increased depletion size. The injected collector current also reduces the capacitance since the injected electron neutralizes the collector depletion donor and extends the depletion layer size. The capacitance variation is shown in Fig. 5.2. Since the current level is larger for the $R_{OPT}$ operation than the $2R_{OPT}$ operation, the feedback capacitance $C_{bc}$ is larger for $2R_{OPT}$ operation with similar $V_{bc}$ swing. Therefore, the gain at $2R_{OPT}$ is reduced significantly by the strong feedback capacitance $C_{bc}$.

Fig. 5.3 shows the simulation results of the power gains for various output loads using the real device model. As shown in the figure, the power gains for both cases with the $R_{OPT}$ and $2R_{OPT}$ are almost the same value regardless of the HBT size. The gain characteristic of the carrier amplifier makes it difficult to get a good AM-AM characteristic from the Doherty amplifier, but still linear Doherty amplifier operation is possible.

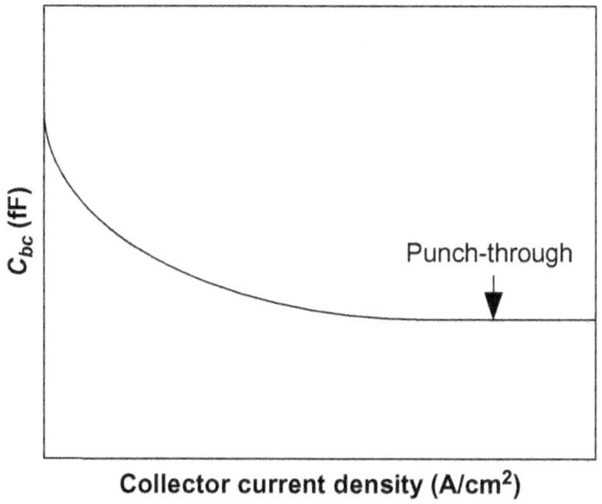


**Fig. 5.2** Extracted $C_{bc}$.

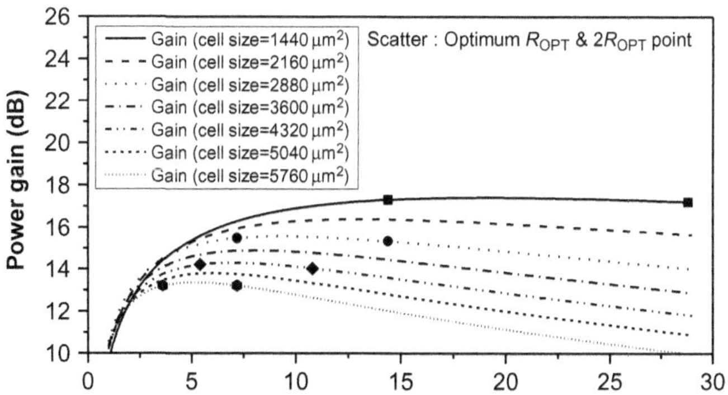


**Fig. 5.3** The simulated power gains versus the load conditions for various transistor cell sizes.

#### *5.2.1.2 Flat Gain Operation of Doherty Amplifier Based on HBT*

Under the ideal condition, the carrier amplifier operation with $2R_{OPT}$ load, at a low power region, has 3 dB higher power gain than the amplifier operation with the $R_{OPT}$, but there is 3 dB input power loss, compensating the gain. At the higher power operation, the gain reduction of the carrier amplifier with the $R_{OPT}$ load is compensated by the peaking amplifier. Therefore, the constant gain is maintained. However, in real device implementation, the gain reduction cannot be compensated by the peaking amplifier because it requires the $g_m$ of the peaking amplifier two times larger than that of the carrier amplifier as described in Section 1.2. It means that the peaking transistor should be two times larger than that of the carrier amplifier, wasting the power generation capability of the peaking amplifier. Moreover, the gain of the peaking amplifier is a lot lower than that of the carrier amplifier due to the class C bias. Therefore, the total gain of the Doherty power amplifier is distorted as shown in Fig. 5.4 (Gain_ideal case). This distortion can be compensated somewhat by the uneven driving as discussed in Section 2.1.

In the real device operation, the gains of the carrier amplifier with the $R_{OPT}$ and $2R_{OPT}$ loads are almost the same value. Therefore, to get the proper Doherty load modulation characteristic, the carrier amplifier should be saturated in the modulation region with about 3 dB gain compression at the maximum output power by driving the carrier amplifier with a higher power. Due to the lower saturated gain of the carrier amplifier, the peaking amplifier can compensate the low gain, assisting for the proper Doherty operation with a constant gain as shown in Fig. 5.4. In the low-power region, only the carrier amplifier operates and should be linear. Therefore, the carrier amplifier should be designed as a linear amplifier at $2R_{OPT}$ load with a class AB bias. At the higher-power region, the IM3 components generated by the carrier amplifier and peaking amplifier should be canceled.

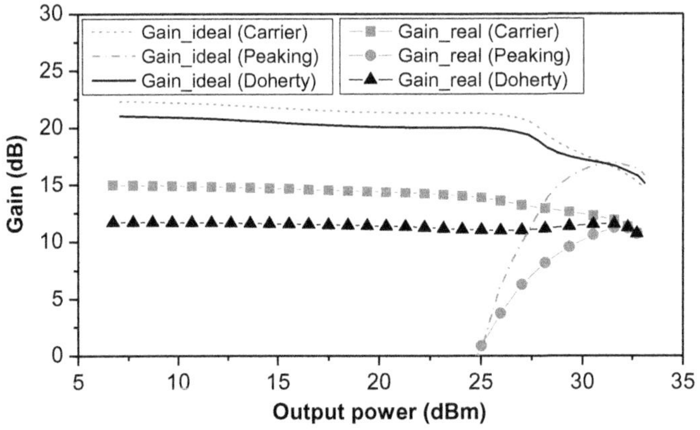


**Fig. 5.4** Simulated gains of the carrier, peaking, and Doherty amplifiers under Doherty operation.

### 5.2.2 IMD3 Cancellation With Proper Harmonic Load Conditions

At the load modulation region with the quasisaturated operation of the carrier amplifier, the cancellation of the third-order intermodulation distortions (IMD3s) generated by the carrier and peaking amplifiers is the most important design issue in realizing a linear Doherty power amplifier. In a simple power series expression of the gain using a two-tone signal, the fundamental output power is expressed as

$$v_{\text{Fund}} = \left\{ a_1 v + \frac{9}{4} a_3 v^3 + \frac{25}{4} a_5 v^5 \right\} \cos\left(\omega_{1,2} t\right) \tag{5.1}$$

The output of the third-order intermodulation (IM3) is given by

$$v_{\text{IM3}} = \left\{ \frac{3}{4} a_3 v^3 + \frac{25}{8} a_5 v^5 \right\} \cos\left(2\omega_{1,2} - \omega_{2,1}\right) t \tag{5.2}$$

Here, $a_1$ is the linear gain coefficient, and the other terms $a_i$ represent the higher-order harmonic generation coefficients as described in Section 4.1.3. In Eq. (5.1), the linear gain is distorted by the higher-order harmonic generation.

The gain expansion occurs for a class C biased amplifier since $a_3/a_1$ and $a_5/a_1$ are positive. Therefore, the IM3 and fundamental signals of the peaking amplifier are in-phase. On the other hand, the gain compression occurs in the class AB biased carrier amplifier with negative $a_3/a_1$ and $a_5/a_1$, and the IM3 and fundamental signals are in antiphase. By using the complementary gain characteristics of the carrier and peaking amplifiers, the IMD3s of the two amplifiers can cancel each other, thereby improving the linearity of the Doherty power amplifier. These coefficients are strongly bias-dependent, and the gate biases of the carrier and peaking amplifiers should be properly adjusted for the cancellation.

In Eq. (5.2), the IMD3s at the lower and higher frequencies of the two-tone signal are the same magnitude, and the symmetrical IMD3 can be canceled. However, the different phased IM3s having different magnitudes are produced by the interactions of the fundamental and the feedback even-order intermodulation harmonic terms. This nondirectly generated IM3 terms, not included in the power series equations, cannot be canceled, deteriorating the IMD3 cancellation. The IM2, which is the most important even harmonic, should be suppressed by a second-harmonic termination, thereby alleviating the IM3 generation.

Fig. 5.5 shows the ideal simulation of the IM2 generation according to the conduction angle. As shown, the class AB and C amplifiers generate significant IM2s with different magnitudes according to the conduction angle. To investigate the harmonic suppression behavior, a two-tone simulation is conducted. In this simulation, an ideal second-harmonic short termination shown in Fig. 5.6 is attached under the condition that it doesn't affect the fundamental and other harmonic loads. The second-harmonic

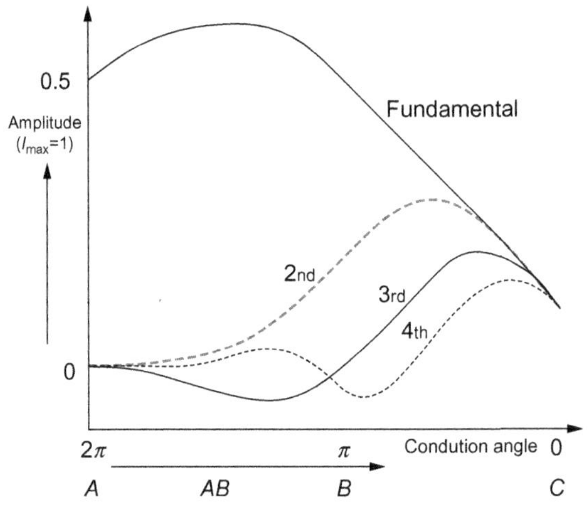


**Fig. 5.5** Fourier analysis for the current waveform of the reduced conduction angle operation.

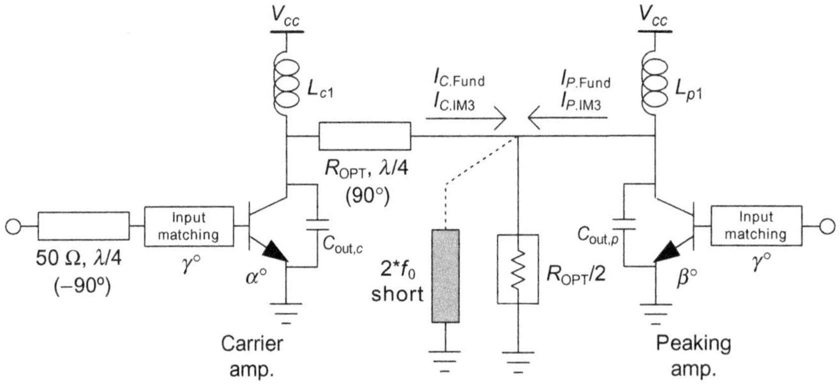


**Fig. 5.6** Doherty power amplifier using a quarter-wave transmission line for the second-harmonic short.

short at the output of peaking amplifier provides the same short condition at the output of the carrier amplifier because the quarter-wave line at a fundamental frequency turns into a half-wave line at the second-harmonic frequency.

Fig. 5.7A shows the magnitude and phase differences between the IM3s of the carrier and peaking amplifiers, $I_{C.IM3}$ and $I_{P.IM3}$. With the second-harmonic short, $I_{C.IM3}$ and $I_{P.IM3}$ are in near antiphase condition because the additional components produced by the even-order harmonics are suppressed. Also, the magnitude difference is in near zero. Therefore, the Doherty power amplifier with the second-harmonic short has a lower IMD3, through the IM3 cancellation, than the other case as shown in Fig. 5.7B. Moreover, the power amplifier with the second-harmonic short delivers a

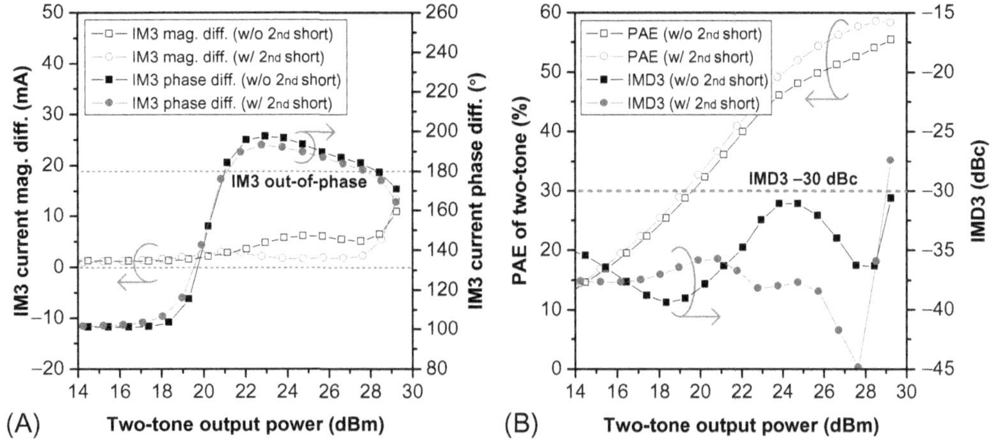


**Fig. 5.7** Two-tone simulation results with and without the second-harmonic short termination at the output: (A) magnitude and phase differences of IM3 currents from the carrier and peaking amplifiers versus the power sweep and (B) PAE and IMD3 curves versus the power sweep.

higher power-added efficiency (PAE) because it produces a class F waveform. The second-harmonic short circuit generates a half-sinusoidal shaped current waveform with the in-phase second-harmonic current, and the high third-harmonic load is easily produced internally for a rectangular-shaped voltage waveform.

The efficiency and linearity at the 6 dB power back-off (PBO) region are closely related to the turn-on timing of the peaking amplifier. Since the peaking amplifier cannot be abruptly turned on, it should be turned on early to compensate the AM-AM and AM-PM distortions of the carrier amplifier for achieving high linearity at the 6 dB PBO, but the early turn-on means that the load modulation starts early, before the carrier reaches to the saturated operation, and the efficiency at the 6 dB PBO is reduced.

Fig. 5.8 shows the effect of the turn-on timing of the peaking amplifier. The lower bias of the peaking amplifier boosts the efficiency of the 6 dB PBO, but the peak efficiency is decreased since the output power of the peaking amplifier is reduced due to the deeper class C bias. Moreover, the AM-AM and AM-PM are deteriorated leading to a poor linearity. For a linear Doherty power amplifier, therefore, the base bias of the peaking amplifier should be chosen properly to make the linear AM-AM and AM-PM responses.

## 5.3 COMPACT DESIGN FOR HANDSET APPLICATION

Compact design of the Doherty amplifier is essential for handset applications. Fig. 5.9A shows a conventional Doherty structure, which uses a power divider such as a coupler for the inputs of the both amplifiers, a quarter-wavelength line for load

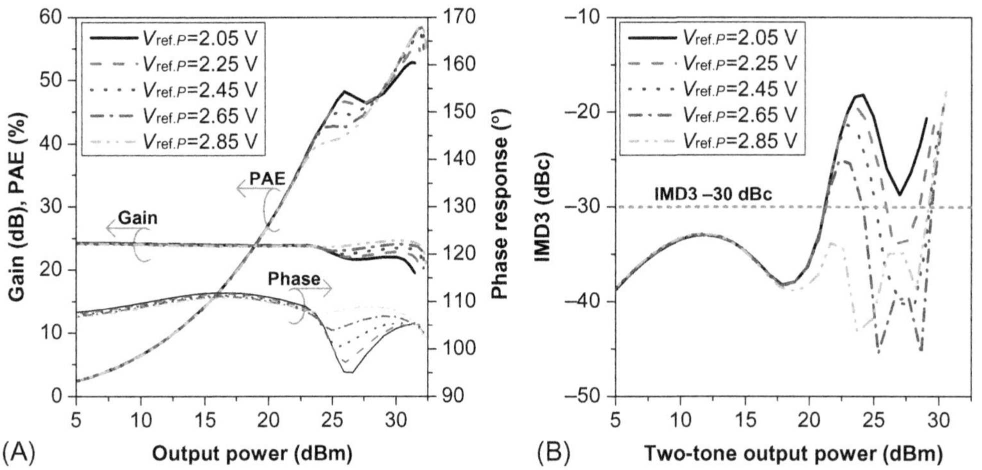


**Fig. 5.8** (A) Variations of the simulated gain and PAE of Doherty amplifier and (B) the two-tone simulated IMD3 with the base bias sweep of the peaking amplifier.

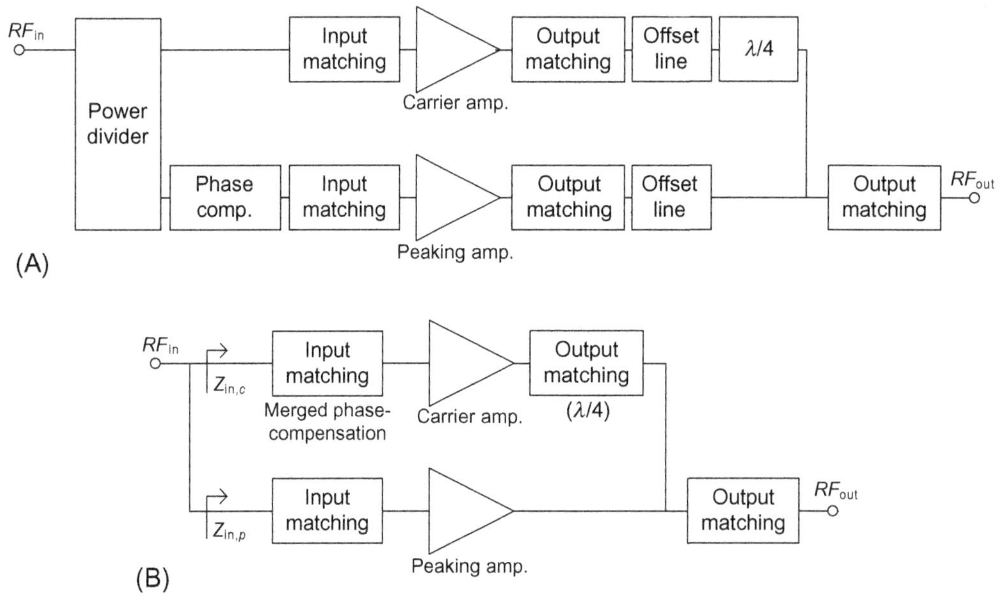


**Fig. 5.9** Block diagrams of the (A) conventional and (B) compact Doherty topologies.

modulation, and offset lines for imaginary impedance modulation. The functional blocks can be simplified by merging the functional circuit components as shown in Fig. 5.9B. The offset lines are eliminated by terminating the output capacitance at the drain terminal before the output matching circuit, thereby eliminating the imaginary part impedance transformation by the matching circuit. The input power divider can be eliminated by the direct power-dividing approach. However, the coupler-based power divider provides a lot of stable operation with accurate power-dividing ratio, and it is a design option between the compact size and the stable operation.

### 5.3.1 Input Power Dividing Circuit

Fig. 5.10 shows the two input-power-dividing circuits: coupler divider and direct power divider. For mobile applications, a bulky input power divider can be removed by a direct input-power-dividing technique shown in Fig. 5.10B. For the ideal input power driving, the carrier amplifier should receive more power in the low-input-power region when the peaking amplifier is turned off, compensating the low gain at $2R_{OPT}$ operation, and the peaking amplifier gets more power in the high-input-power region to increase the gain and output power of the class C biased peaking amplifier.

The uneven power dividing can be realized using the power-level-dependent input impedance variations of the carrier and peaking amplifiers. The input capacitance $C_{in}$ of an HBT shown in Fig. 5.1 is determined by $C_{be}$ and miller capacitance $C_{bc}$. The capacitance is expressed as.

$$C_{in} = C_{be} + C_{bc}(1 + g_m(R_O \| R_L)) \text{ (where } R_O \gg R_L) \tag{5.3}$$

where $R_L$ is the output load impedance. Fig. 5.11 shows the $C_{in}$ variations for the carrier and peaking amplifiers.

Because of the class C bias of the peaking amplifier, the large-signal transconductance ($g_m$) and $C_{be}$ of the peaking amplifier vary significantly, while those of the carrier amplifier remain almost constant. The input capacitance of the peaking amplifier increases over 200% as the input power increases, while that of the carrier amplifier varies less than 10%. The power-level-dependent input impedances can be matched to 50 Ω at a high-power region using either low-pass filter (LPF) or high-pass filter (HPF) type circuit as shown in Fig. 5.12. The carrier amplifier could be matched to 50 Ω for all power levels, since its impedance variation is small. For the same matching, the peaking amplifier shows significant mismatches at a low-power region.

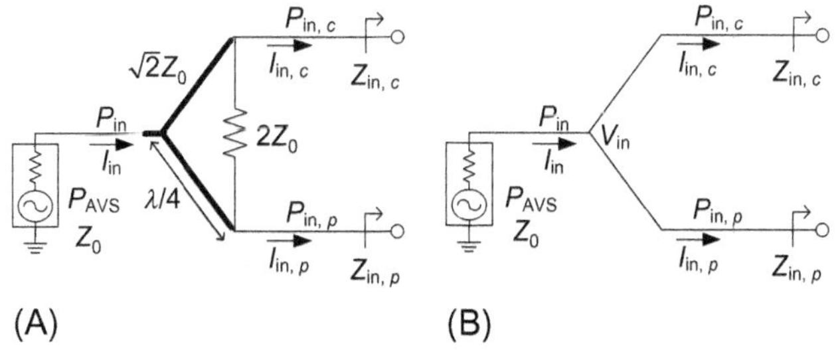


**Fig. 5.10** (A) Wilkinson power divider for Doherty amplifier. Two output ports are isolated. (B) Direct input dividing circuit.

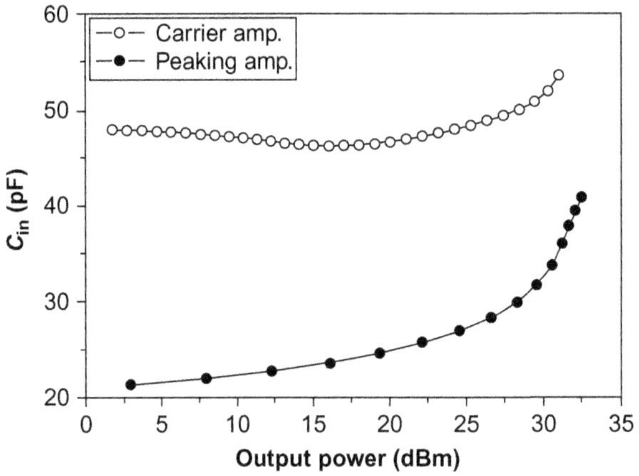


**Fig. 5.11** Input capacitance variations of the carrier and peaking amplifiers.

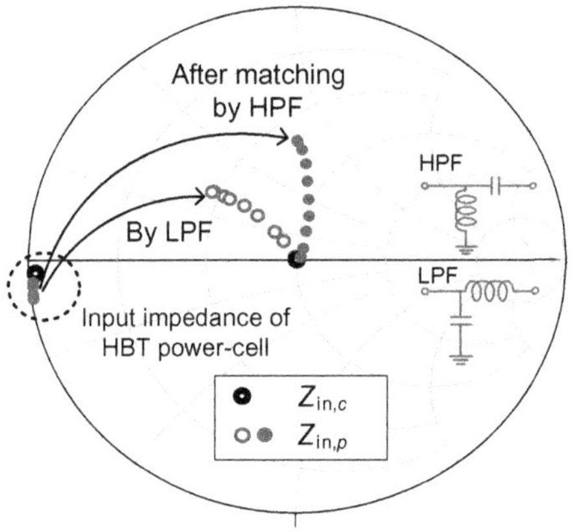


**Fig. 5.12** Simulated input impedances of a HBT power cell and the impedance traces varied with the increased input power after matching to 50 Ω by LPF- and HPF-type circuits.

As depicted in Fig. 5.12, the LPF-type circuit converts the input admittance of the peaking to higher than 1/50 ℧ at a low-power operation, while the HPF-type circuit converts it to lower than 1/50 ℧. The trace in Fig. 5.12 with the HPF matching indicates that the carrier amplifier receives more power at a low-input-power region because of its higher admittance, and the peaking amplifier can get more power at a high-input-power region. This HPF-type matching can provide the desired uneven input drive.

#### *5.3.1.1 Input Power Dividing Using a Coupler*

Wilkinson power dividers or 90° hybrid couplers are normally used for the input divider of a Doherty amplifier. With the power divider shown in Fig. 5.10A, the ports of the carrier and peaking amplifiers are isolated, and the input drive powers to the carrier and peaking amplifiers are determined by the coupler. The coupled powers to the amplifiers are related to the reflection coefficients ($S_{11}$) of the carrier and peaking amplifiers. The coupled power to the carrier amplifier is given by

$$\begin{aligned} P_{\text{in},c} &= \frac{1}{2} P_{\text{AVS}} \left(1 - |S_{11}|^2\right) \\ &= \frac{1}{2} P_{\text{AVS}} \left(1 - \left|\frac{Z_{\text{in},c} - Z_0}{Z_{\text{in},c} + Z_0}\right|^2\right) \end{aligned} \tag{5.4}$$

The coupled power to the peaking amplifier is the same as Eq. (5.4) but is determined by $Z_{\text{in},p}$. The coupled power contours to the ports according to $Z_{\text{in},c}$ and $Z_{\text{in},p}$ are depicted in Fig. 5.13A. Because of the isolation property, the powers delivered to the two ports are independent.

If the input impedances of the carrier and peaking amplifiers are matched to $Z_0$ at the maximum output power using the HPF-type circuit and the input is divided by the coupler, the input power is driven in the two ports as shown in Fig. 5.14A. Because $Z_{\text{in},p}$ is mismatched at a low-power operation region due to the impedance variation as depicted in Fig. 5.12, the power delivered to the peaking amplifier at the power region is lower than that at the high-power region. The carrier amplifier is matched always, and the input power is constant. This input drive is similar to the even input drive and is not the optimum condition for the Doherty operation. For the proper Doherty amplifier operation, the input power driving should be an uneven drive as shown in Fig. 5.14B. When the input impedance of the peaking amplifier is matched to $Z_0$ and that of carrier amplifier to −3.5 dB circle of Fig. 5.13A, the carrier input power is reduced by 1 dB, and 1 dB more power is delivered to the peaking amplifier at the maximum output power, realizing the uneven drive for optimum operation of the Doherty amplifier. Of course, this uneven drive can be realized using a coupler with the proper coupling ratio instead of mismatch at the carrier amplifier as described in Section 2.1.

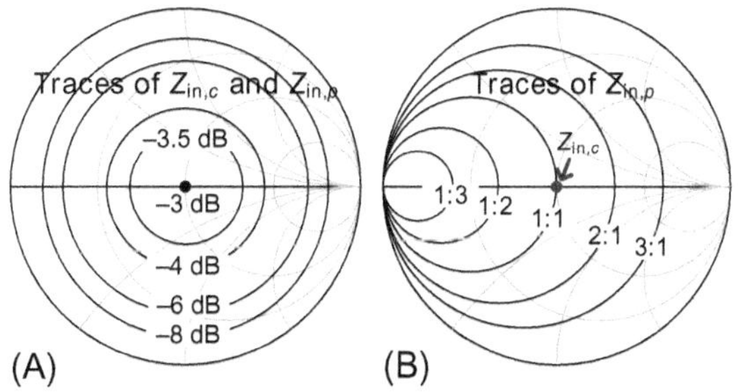


**Fig. 5.13** (A) Coupled power contours ($P_{\text{in},c}/P_{\text{in},p}$) to the ports using a coupler. $P_{\text{in},c}$ and $P_{\text{in},p}$ are independent. (B) Power dividing ratio contours according to the trace of $Z_{\text{in},p}$ with the direct input dividing. $Z_{\text{in},c}$ is matched to $Z_0$.

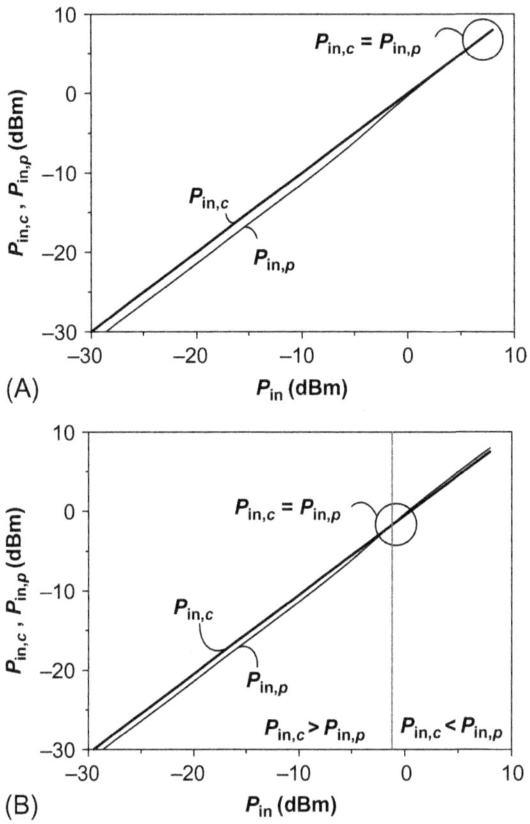


**Fig. 5.14** (A) Input power driving with a coupler, when the input impedances of carrier and peaking amplifiers are matched to $Z_0$ at the maximum output power. (B) Uneven power driving, which can be achieved by the 3 dB coupler.

### *5.3.1.2 Direct Input Dividing Without Coupler*

A direct input dividing can be realized using the impedance variations of the carrier and peaking amplifiers. Unlike the power dividers, the direct input power divider does not have isolation between the carrier and peaking amplifier inputs since the carrier and peaking amplifiers share the same voltage node at the input as shown in Fig. 5.10B.

When the input impedances of the amplifiers are assumed to be $Z_{\text{in},c}$ and $Z_{\text{in},p}$, respectively, the power at the junction is given by

$$\begin{aligned} P_{\text{in}} &= P_{\text{AVS}}\left(1 - |S_{11}|^2\right) \\ &= P_{\text{AVS}}\left(1 - \left|\frac{Z_{\text{in},c}||Z_{\text{in},p} - Z_0}{Z_{\text{in},c}||Z_{\text{in},p} + Z_0}\right|^2\right) \end{aligned} \tag{5.5}$$

Here, $P_{AVS}$ is the source power. At the node, the input voltage is identical, but the input current is divided to the carrier amplifier and the peaking amplifier by the ratio of their admittances. The divided currents and the voltage at the junction determine the input powers to the carrier and peaking amplifiers, respectively. The divided powers are given by

$$P_{\text{in},c} = \frac{1}{2} Re\left[V_{\text{in}} \cdot I_{\text{in},c}{}^{*}\right] = \frac{1}{2} Re\left[\frac{|V_{\text{in}}|^2}{Z_{\text{in},c}{}^{*}}\right] \tag{5.6}$$

$$P_{\text{in},p} = \frac{1}{2} Re\left[V_{\text{in}} \cdot I_{\text{in},p}{}^{*}\right] = \frac{1}{2} Re\left[\frac{|V_{in}|^2}{Z_{in,p}{}^{*}}\right] \tag{5.7}$$

The matched impedances of the carrier and peaking amplifiers are given by

$$Z_{\text{in},c} = R_{\text{in},c} + jX_{\text{in},c}, \quad Z_{\text{in},p} = R_{\text{in},p} + jX_{\text{in},p} \tag{5.8}$$

The power-dividing ratio is equal to the ratio of the input admittances of the amplifiers:

$$\begin{aligned} P_{\text{in},c} : P_{\text{in},p} &= \frac{1}{2} Re\left[\frac{|V_{\text{in}}|^2}{Z_{\text{in},c}{}^{*}}\right] : \frac{1}{2} Re\left[\frac{|V_{\text{in}}|^2}{Z_{\text{in},p}{}^{*}}\right] \\ &= \frac{R_{\text{in},c}}{R_{\text{in},c}{}^2 + X_{\text{in},c}{}^2} : \frac{R_{\text{in},p}}{R_{\text{in},p}{}^2 + X_{\text{in},p}{}^2} \end{aligned} \tag{5.9}$$

The input impedance of the carrier amplifier remains almost constant for the input power variation, while that of the peaking amplifier changes significantly because of the class C bias. For 1: $N$ power dividing, with the matched carrier amplifier, the transformed input impedance of the peaking amplifier is given by

$$1 : N = 1 : \frac{r_{\text{in},p}}{r_{\text{in},p}{}^2 + x_{\text{in},p}{}^2} \tag{5.10}$$

where $r_{\text{in}}$ and $x_{\text{in}}$ are normalized by $Z_0$. Eq. (5.10) becomes

$$\left(r_{\text{in},p} - \frac{1}{2N}\right)^2 + x_{\text{in},p}{}^2 = \left(\frac{1}{2N}\right)^2 \tag{5.11}$$

which is an equation of a circle centered at $r = 1/(2N)$ and $x = 0$, with radius of $1/(2N)$, and is represented by a conductance circle on the Smith chart. In the case of $N = 1$, $r_{\text{in},p}$ and $x_{\text{in},p}$ are on the $Z_0$ conductance circle on the Smith chart. Fig. 5.13B shows a power-dividing ratio contours according to the trace of $Z_{\text{in},p}$, where $Z_{\text{in},c}$ is matched to $Z_0$. Fig. 5.15 shows the input impedances of the peaking amplifier for 1:1 dividing for the even drive and 1:2 dividing for the uneven drive. When $Z_{\text{in},p}$ is located inside of the 1:1 circle, the uneven drive is realized, delivering more power to the peaking amplifier. Following $Z_{\text{in},p}$ matched by the HPF circuit shown in Fig. 5.12, the trace of the uneven

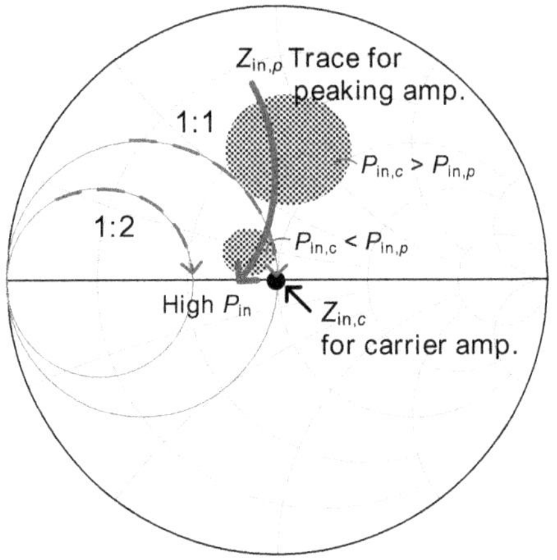


**Fig. 5.15** 1:1 and 1:2 uneven input power dividings determined by the input impedances. Impedances are normalized by $Z_0$.

$P_{in}$ drive is depicted in Fig. 5.15. The trace indicates that the carrier amplifier receives more power at a low-input-power region and the peaking amplifier gets more power at a high-input-power region, which is the desired input drive.

### *5.3.1.3 Realization of the Input Power Dividing Circuit*

The phase compensation network of the quarter-wavelength line at the input of the peaking amplifier as shown in Fig. 5.9A can perturb the proper input power dividing. Alternatively, the phase compensation network is employed at the input of the carrier amplifier using a HPF-type circuit with negative angle as shown in Fig. 5.16. A quarter-wavelength line for the phase compensation is implemented in a lumped

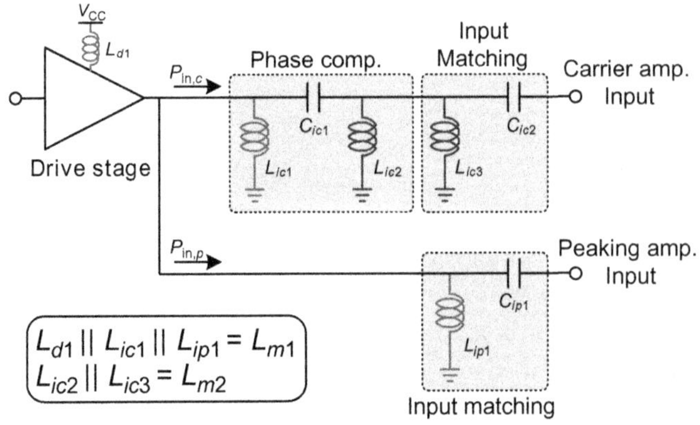


**Fig. 5.16** Schematic of the input matching circuit for the Doherty stage with the direct input-power-dividing technique. The circuit configuration is simplified by merging the inductors.

$\pi$-type network using two shunt inductors considering the rearrangement. The amplifiers are matched using the HPF-type matching circuit for the uneven drive. The five inductors shown in Fig. 5.16 can be merged into two inductors, and both the number and size of the inductors are reduced owing to the parallel structure. $L_{m1}$ and $L_{m2}$ are easily implemented by a collector bias line of the drive stage and a bond wire, respectively. Overall, two capacitors ($C_{ip1}$ and $C_{ic2}$) and one $\pi$-circuit are needed to drive the Doherty stage along with the interstage matching.

## 5.3.2 Output Circuit Implementation

Fig. 5.6 shows a Doherty amplifier using a quarter-wave inverter. The output capacitances of the carrier ($C_{out,c}$) and peaking ($C_{out,p}$) are resonated out at the drain by the inductances of the bias lines $L_{c1}$ and $L_{p1}$, respectively. Therefore, the matching circuits do not need to transfer the imaginary part of impedance, and the offset lines are not needed, achieving a smaller size circuit. The phase compensation circuit at the input of the carrier amplifier and the quarter-wave inverter at the output have phase delays of $-90°$ and $90°$, respectively, and the phase delay between the two amplifiers are adjusted.

To implement the amplifier on a single chip for a handset application, the bulky quarter-wave inverter should be replaced by an equivalent lumped LC network as shown in Fig. 5.17. The lumped network type should be chosen carefully by considering the compact design and the harmonic load condition. In the size aspect, (A) and (B) types can share a single collector bias line, but (C) and (D) require two collector bias lines due to the DC blocking capacitor. The shunt inductors of (C) can be used for collector bias lines of the carrier and peaking amplifiers, respectively, but the inductance should be large to compensate the output capacitors $C_{out,c}$ and $C_{out,p}$.

With the high-pass $\pi$ type (C), both the number and values of the lumped components can be significantly reduced, similarly to the elaborated matching technique at the input that is explained in Section 5.3.1.3. However, (A) and (B) circuits have a size advantage at the output. Since the shunt capacitors of the low-pass $\pi$ type (A) can be merged with the output capacitors $C_{out,c}$ and $C_{out,p}$, their values and sizes are reduced. In general, an output matching circuit of a power amplifier for a handset is implemented by off-chip components on the package module to reduce the output matching circuit

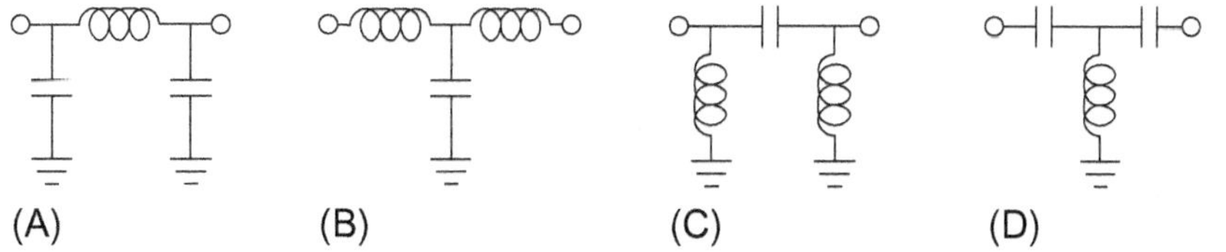


**Fig. 5.17** Equivalent lumped quarter-wave inverters: (A) a low-pass $\pi$ type (90°), (B) a low-pass T type (90°), (C) a high-pass $\pi$ type (−90°), and (D) a high-pass T type (−90°).

loss. Due to the low characteristic impedance of $R_{OPT}$ at ~GHz band, the inductance of the lumped inverter is lower than 1 nH (for $R_{OPT} = 8\ \Omega$ device, $L_{c1} = 0.67$ nH). Therefore, a series inductor of (A) can be implemented by a bond-wire inductance that is employed to connect the chip to a package module.

The lumped inverter of (A) is the most suited structure to reduce the size of the network, but the size of the harmonic load should also be considered. To analyze the circuit in detail, harmonic balance simulation with the schematic shown in Fig. 5.18 is conducted, and the results are shown in Fig. 5.19. With the lumped inverter, the second-harmonic shorts at the outputs of the carrier and peaking amplifiers have a minor effect on linearity since the large shunt capacitors $C_{c1}$ and $C_{c2}$ provide near short second-harmonic load conditions for the carrier and peaking amplifiers, respectively. Therefore, the amplifier can deliver a good IMD3 characteristic without the second-harmonic shorts.

However, the second-harmonic shorts at the outputs of the carrier and peaking amplifiers show different effects on the PAE. When the second-harmonic short is provided at the carrier amplifier, it reduces the PAE because the out-of-phase second-harmonic current is increased and bifurcated current waveforms is generated, reducing the ratio of $I_{c.\mathrm{Carrier.Fund}}/I_{c.\mathrm{Carrier.DC}}$ as shown in Fig. 5.20A. On the other hand, when the second-harmonic short is provided at the peaking amplifier, it improves the PAE because the in-phase second-harmonic current is increased and the half-sinusoidal peaked current waveform is generated, increasing the ratio of $I_{c.\mathrm{Peaking.Fund}}/I_{c.\mathrm{Peaking.DC}}$ as shown in Fig. 5.20B. Consequently, the peaking amplifier with the second-harmonic short delivers a higher PAE in contrast to the carrier amplifier. Therefore, the short circuit is employed only at the peaking amplifier.

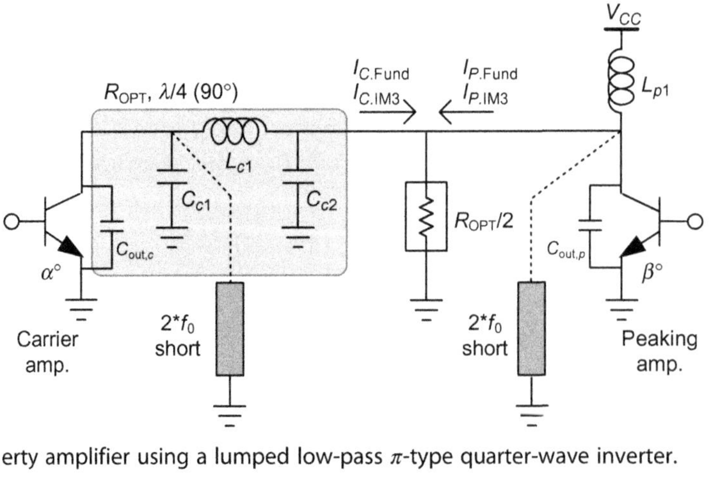


**Fig. 5.18** Doherty amplifier using a lumped low-pass π-type quarter-wave inverter.

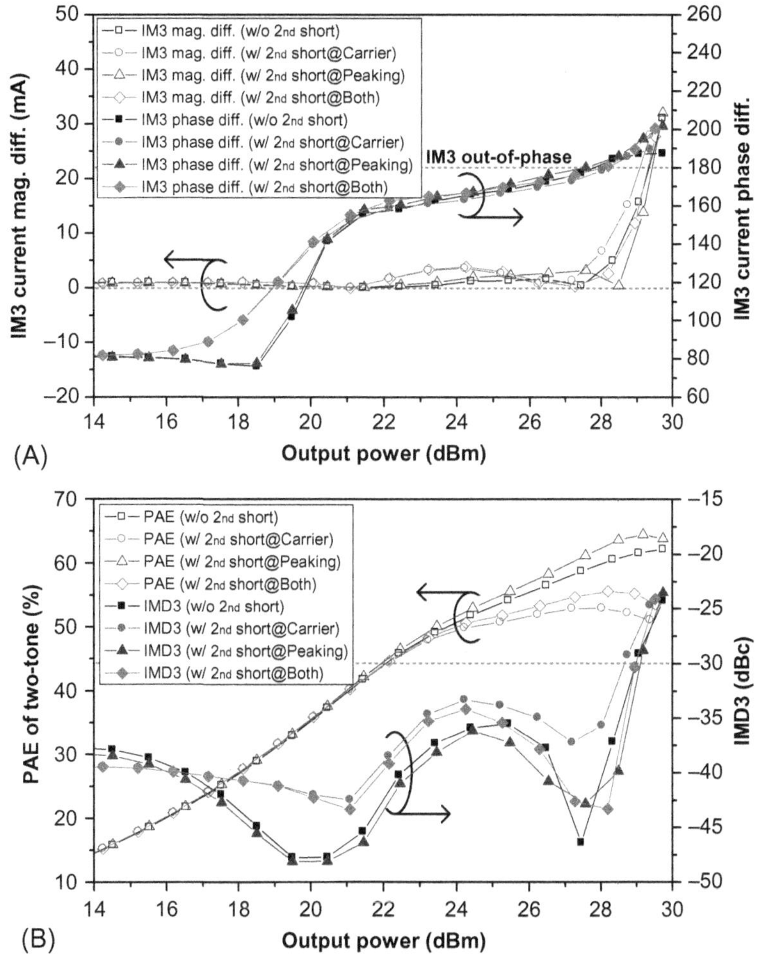


**Fig. 5.19** Two-tone simulation results with and without the second-harmonic short terminations at the output: (A) magnitude and phase differences of IM3 currents between the carrier and peaking amplifiers versus the power sweep and (B) PAE and IMD3 curves versus the power sweep.

## 5.4 IMPLEMENTATION AND MEASUREMENT

Total schematic of the linear Doherty power amplifier and its implemented chip photograph are shown in Fig. 5.21. The power amplifier is fabricated using an InGaP/GaAs HBT, and the chip area is 1.1 × 1.2 $mm^2$. With a supply voltage of 3.4 V, the quiescent currents of the amplifier are 24 and 36 mA for the drive and power stages, respectively. Fig. 5.22 shows the measured PAE, gain, and IMD3 of the power amplifier for a 10 MHz tone-spacing two-tone signal at 1.9 GHz. The power amplifier delivers not only

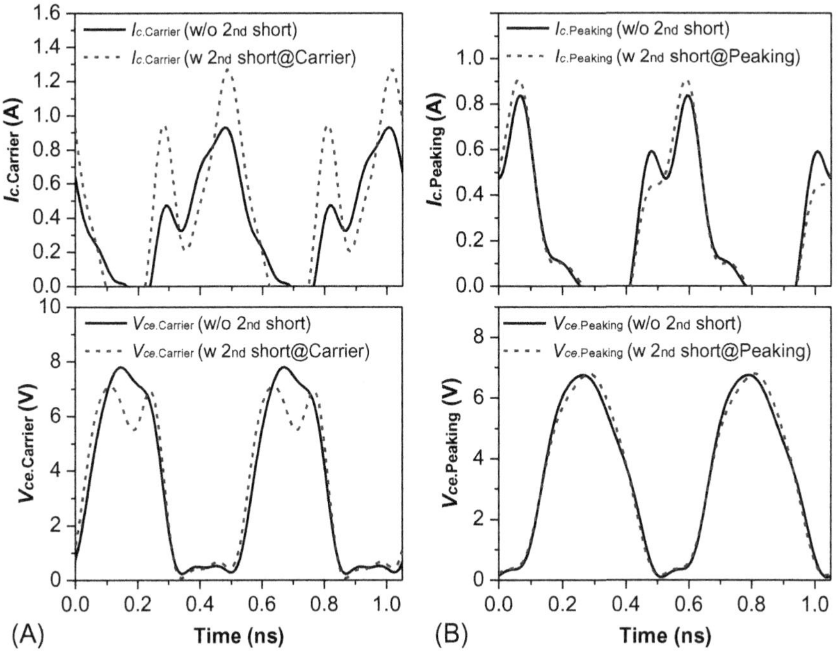


**Fig. 5.20** Simulated output collector current $I_c$ and collector-emitter voltage waveforms ($V_{ce}$) without and with the second-harmonic short termination at the output of (A) carrier amplifier and (B) peaking amplifier, respectively.

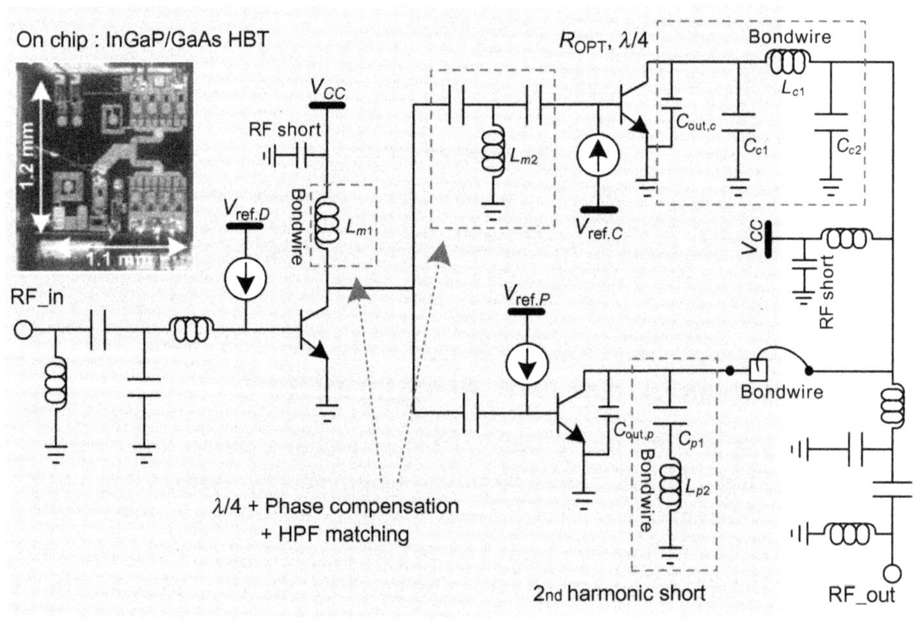


**Fig. 5.21** Schematic of the two-stage linear Doherty power amplifier and the fabricated chip photograph.

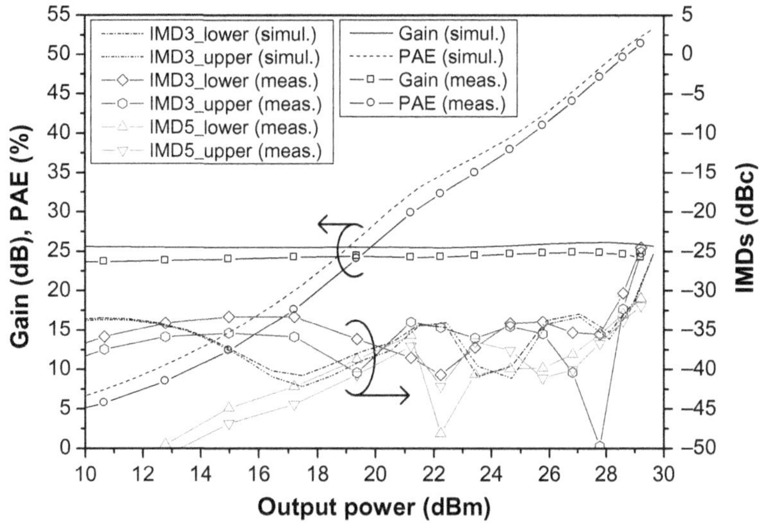


**Fig. 5.22** Measured performances of the power amplifier with a 10 MHz tone-spacing two-tone signal. The gain, PAE, IMD3, and IMD5 performances versus the power sweep.

good linearity but also high efficiency by using the proper harmonic control as described before. For a 10 MHz BW 16-quadrature amplitude modulation (QAM) 7.5 dB PAPR LTE signal at 1.9 GHz frequency, the power amplifier delivers a gain of 25.1 dB, a PAE of 45.8%, and an evolved universal terrestrial radio access adjacent channel leakage ratio (E-UTRA$_{ACLR}$) of −35 dBc at an average output power of 27.5 dBm as shown in Fig. 5.23. Without any additional chip like an ET amplifier, this simple Doherty power amplifier delivers the high performance at the peak output power and at the PBO regions.

## 5.5 DOHERTY POWER AMPLIFIER BASED ON CMOS PROCESS

In CMOS power amplifier, a differential structure based on a transformer is widely used because this architecture is well suited to solve the low breakdown voltage and low power density of a CMOS device. In this architecture, the differential output power is voltage-combined, increasing the output load impedance and providing a good virtual ground for a large transistor with a low impedance. Therefore, a voltage-combined Doherty power amplifier based on the transformer is a natural choice for the CMOS process. As shown in Section 1.4.1, the series load can be properly modulated with the output transformer. The series Doherty amplifier modulates the load exactly the same way with the conventional current combining Doherty amplifier. The only difference is the output powers from the carrier and peaking amplifiers are combined in voltage way.

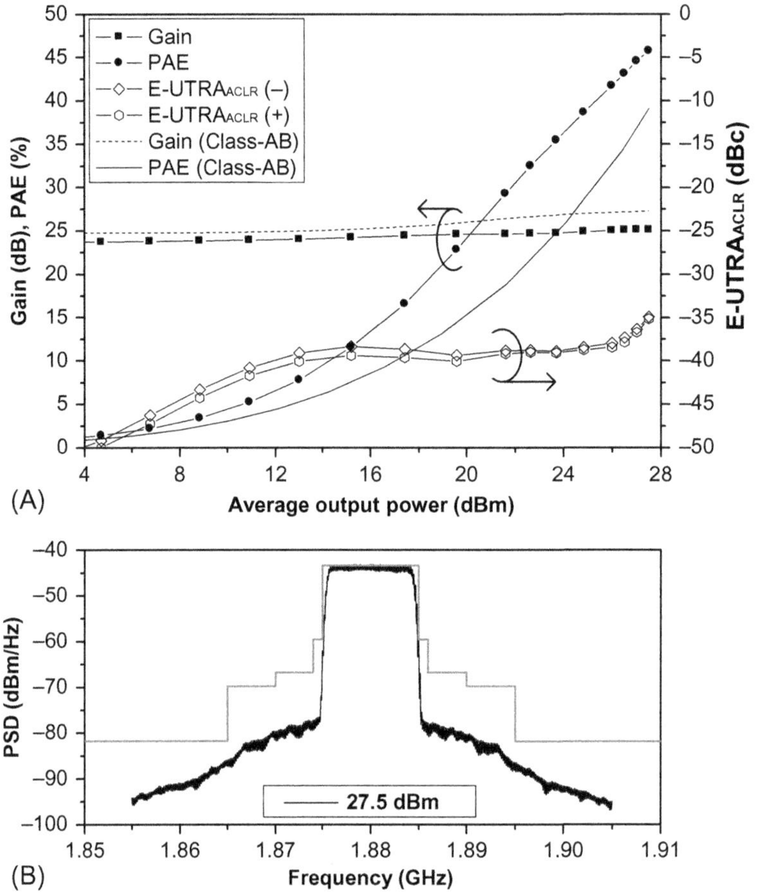


**Fig. 5.23** (A) Measured gain, PAE, and E-UTRA$_{ACLR}$ performances versus the power sweep for the LTE signal. (B) Output spectrum at 27.5 dBm output.

### 5.5.1 Implementation of Linear CMOS Doherty Power Amplifier

The schematic of the CMOS Doherty power amplifier including the output transformer is shown in Fig. 5.24. The carrier and peaking amplifiers are cascode amplifiers, and they are voltage-combined using the transformer. Thick-oxide 0.4 μm gate transistors are used in the common-gate (CG) stages for reliable high-power operation, and thin-oxide 0.18 μm gate transistors are used in the common-source (CS) stages to enhance transconductance and gain. The total gate widths of the thin and thick devices are 4800 and 9600 μm, respectively. The second-harmonic short circuits implemented by metal-insulator-metal (MIM) capacitors and the inductances of a down-bonding wires are provided at the sources of the CS stages to improve linearity by providing the proper second-harmonic grounding and suppressing the second-harmonic term as described before. The $R_{CG}$-$C_{CG}$ bias circuits are inserted to control the voltage swings at the gates of the CG stages. The $L_{in,c}$ and $L_{in,p}$ of the

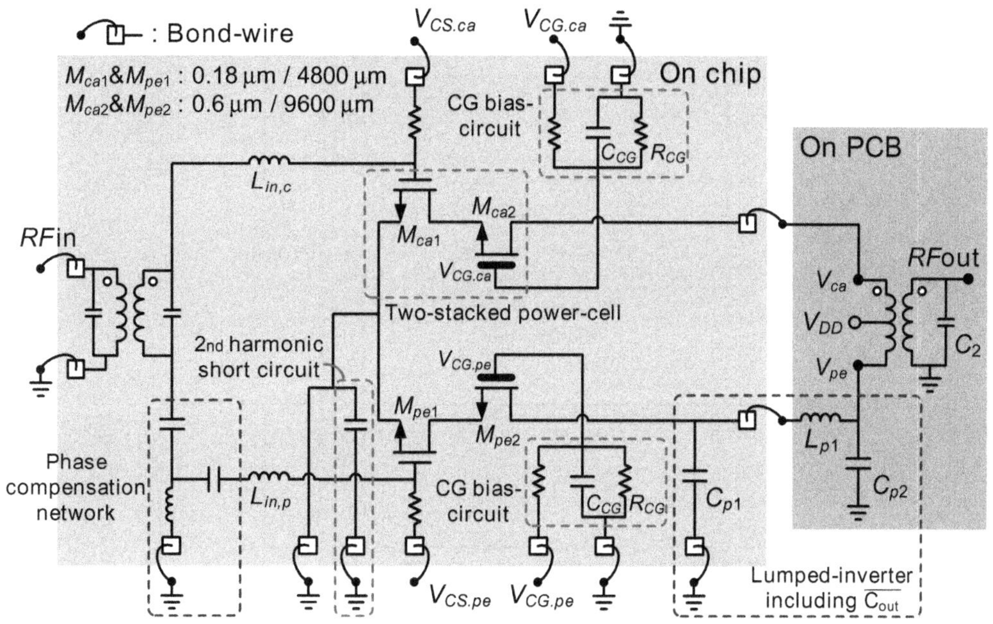


**Fig. 5.24** Schematic of a voltage-combined CMOS Doherty power amplifier based on a transformer.

input matching networks transform the input impedances of the transistors to pure resistance of 12 Ω. A 90° high-pass T-type lumped quarter-wave transformer is inserted at the input path of the peaking amplifier to compensate the phase of the output quarter-wave inverter. A transformer is used for the balanced input drive.

At the output, the output capacitances of the devices are de-embedded in the Doherty network components without using the offset line similarly to the current combining case. Fig. 5.25A shows the output matching network of the practical CMOS Doherty power amplifier, and Fig. 5.25B illustrates how to manipulate the output capacitances of the transistors without using the offset lines.

The output capacitance of the peaking transistor, $\overline{C_{out..p}}$ is merged into the quarter-wave inverter, a low-pass $\pi$ type, considering the compact implementation. Thus, the first shunt capacitor ($C_{p1}$) is the lumped inverter capacitance subtracted by the $\overline{C_{out.p}}$. The output transformer with $C_1$ and $C_2$ provides impedance matching for the output. Since the center tap of the output transformer is used as a biasing point for compact implementation, the output capacitance of the carrier amplifier $\overline{C_{out.c}}$ should be $2C_1$ for a simple merged structure without offset line for the carrier amplifier. The $C_1$ and $C_{inverter}$ determine the second shunt capacitor ($C_{p2}$) of the lumped inverter as shown in Fig. 5.25B. A printed-circuit-board (PCB) transformer is designed using two-layer metal lines. The use of the two metal layers generates stronger coupling and has an advantage in size. The $L_{p1}$ and $C_{p2}$, which are parts of the lumped inverter, are also implemented on the PCB using the metal line and the external capacitor, respectively.

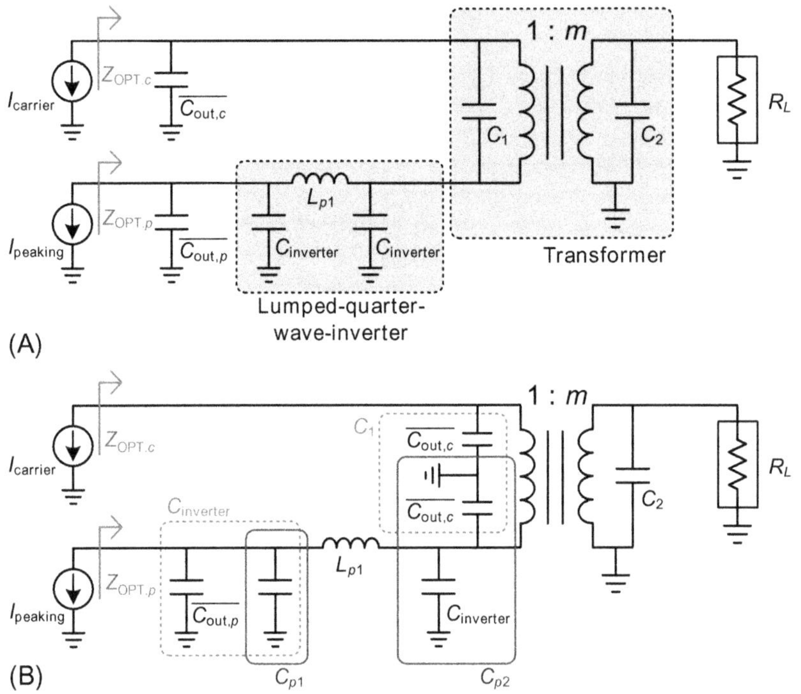


**Fig. 5.25** (A) Practical output matching network with the transmission-line transformer, including $C_1$ and $C_2$. (B) Manipulation of the output capacitances without additional offset lines for the voltage-combined Doherty power amplifier.

The external capacitor $C_2$ is used for the impedance matching with the small inductance of the secondary trace. The insertion loss of the output transformer including the $C_2$ is 0.32 dB at 880 MHz.

## 5.5.2 Measurement Results

The power amplifier is fabricated using a 0.18 μm CMOS process, and the total chip area including the pads is $1.2 \times 1.1$ mm$^2$ as shown in Fig. 5.26. The total amplifier module including the output combining transformer on an FR4 PCB board is $5.0 \times 2.9$ mm$^2$. Fig. 5.27 shows the measured two-tone performance of the power amplifier with a supply voltage of 4.0 V at the 880 MHz. The first peak efficiency at the 6 dB PBO can be maximized by adjusting the CS gate bias voltage of the peaking amplifier. However, the amplifier at the gate bias with the maximized efficiency produces a gain distortion, deteriorating the linearity as shown in Fig. 5.27A. For the linear Doherty power amplifier, the CS gate bias of the peaking amplifier is chosen to make a linear AM-AM and AM-PM characteristics. With the proper CS gate biases and the additional phase delay line merged into the input phase compensation network, the Doherty power amplifier achieves low IMDs as shown in Fig. 5.27B.

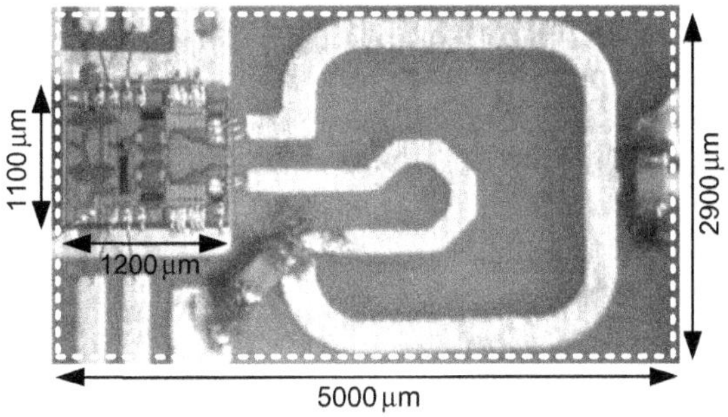


**Fig. 5.26** Photograph of the voltage-combined CMOS Doherty power amplifier module that includes the power amplifier chip in a 0.18 μm CMOS process.

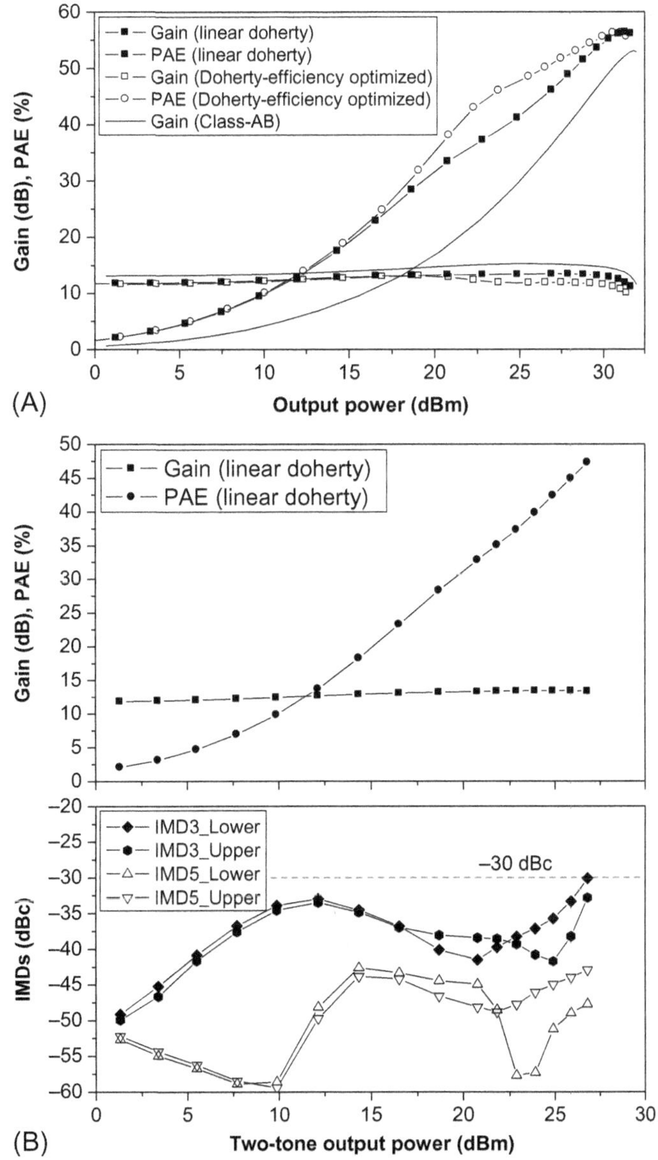


**Fig. 5.27** Measured performance of the Doherty power amplifier with 4.0 V drain bias. The gain, PAE, and IMD performances versus the power sweep for (A) the CW signal and (B) the two-tone signal.

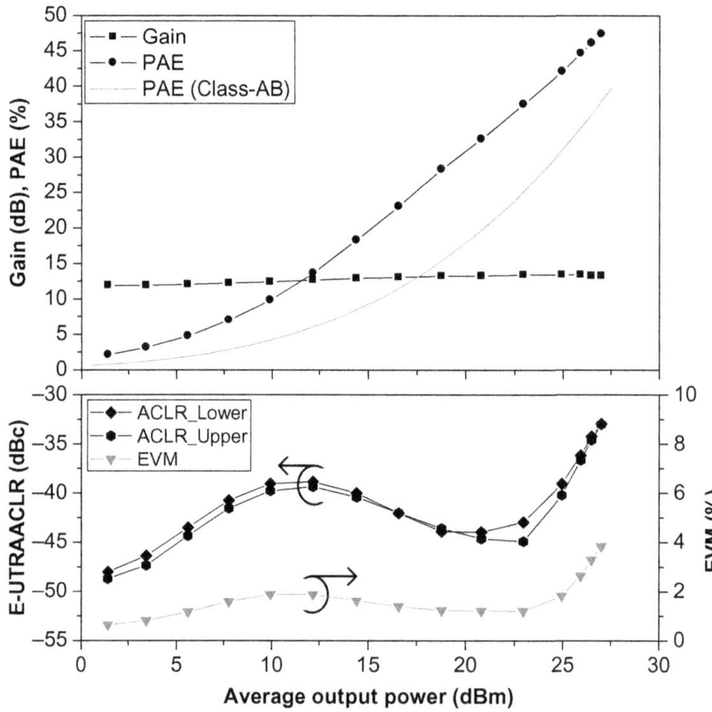


**Fig. 5.28** Measured performance of the power amplifier with the LTE signal at 880 MHz. The gain, PAE, EVM, and E-UTRA$_{ACLR}$ performances versus the power sweep.

The linear power amplifier is tested at 880 MHz using a 10 MHz BW, 16-quadrature amplitude modulation (QAM) and 7.5 dB PAPR LTE signal, and the results are shown in Fig. 5.28. The power amplifier delivers a gain of 13.3 dB, a PAE of 47.4%, an error vector magnitude (EVM) of 3.85%, and an E-UTRA$_{ACLR}$ of −33 dBc at an average output power of 27 dBm. Without using any complex circuitry, the CMOS Doherty power amplifier based on a voltage-combining method could deliver the very good performance for the LTE signal with large PAPR.

## 5.6 AVERAGE POWER TRACKING OPERATION OF THE HANDSET DOHERTY AMPLIFIER

The power amplifier for a handset application should operate efficiently over a large dynamic range, since the power requirement varies with the distance between a base station and a user equipment. Even for the base-station power amplifier, the power control is needed according to the user traffic as described in Section 3.5. Due to the recent trend of multimode operation, one power amplifier should cover the 3G and 4G modulation standards. By referring to the statistical probability density function

(PDF) of the output power for the 3G wideband code division multiple access (WCDMA) system, the probability of transmission at higher than 17 dBm is statistically below 2%, and the power amplifier mostly operates at a back-off power. Also, for the 4G long-term evolution (LTE), the power amplifier spends about 50% of its duration at the back-off power level. As the average output power is decreased, the Doherty power amplifier delivers still a low efficiency. Therefore, the Doherty power amplifier needs to be reconfigured properly for the variation of the average output power. For the purpose, a linear power amplifier with average power tracking (APT) is popularly employed in the handset power amplifier and a similar technology can be applied to the Doherty power amplifier.

For the APT operation of a Doherty amplifier, a multilevel DC-DC converter should be linked to all amplifiers including the carrier, peaking, and drive amplifiers. The cancellation of the IM3s between the carrier and peaking amplifiers is the most important design issue in realization of a linear Doherty power amplifier with APT. However, the IM3 cancellation couldn't be maintained through the various collector biases, deteriorating the linearity. To solve the problem, the base biases of the carrier and peaking amplifiers should be adapted properly to achieve a high linearity for the collector bias sweep.

### 5.6.1 Adaptive Base Bias Control Circuit for Average Power Tracking

Fig. 5.29A shows the schematic of the two-stage Doherty power amplifier. A direct input-power-dividing technique is employed to optimize the load modulation through a proper input drive and to reduce number of circuit components at the input. Flat AM-AM and AM-PM responses are achieved at all power levels by analysis of the gain modulation and selection of proper base biases. To improve the linearity and reduce the size, the lumped quarter-wave transformer of a low-pass $\pi$ type is selected at the output of the carrier amplifier, and the second-harmonic short circuit is employed at the peaking amplifier only.

Fig. 5.30 shows the measurement results with a two-tone signal at 1.9 GHz having 10 MHz tone-spacing. The Doherty power amplifier delivers a good linearity under −30 dBc with 3.4 V $V_{cc}$. However, the linearity is worsen as $V_{cc}$ is decreased. With the voltage lower than 2.4 V $V_{cc}$, the power amplifier does not meet the −30 dBc linearity specification. This means that the IM3 can be cancelled optimally only at the fixed $V_{cc}$, and the circuit should be retuned for proper IM3 cancellation at the different $V_{cc}$.

Since the load is fixed, as the collector bias of the carrier and peaking amplifiers is decreased, the peak output power of the Doherty power amplifier is decreased. Therefore, the input power should be reduced as the peak output power is decreased. Therefore, the carrier amplifier is operated in the low-current region with the low base current, and the base bias of the carrier amplifier should be reduced. To maintain the good linearity with the $V_{cc}$ variation, the turn-on time of the peaking amplifier should be properly

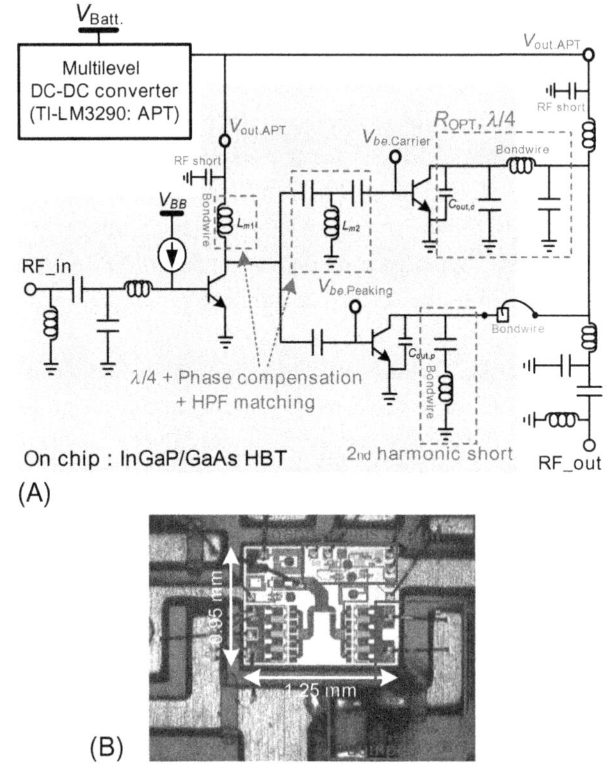


**Fig. 5.29** (A) Schematic of the Doherty power amplifier with multilevel DC-DC converter and (B) fabricated chip photograph.

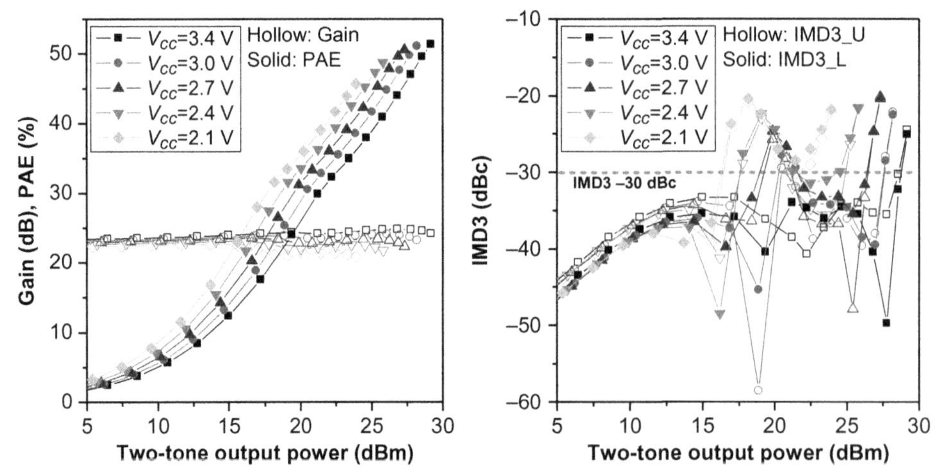


**Fig. 5.30** Measured performances of the Doherty power amplifier with the various collector biases ($V_{cc}$). The gain, PAE, and IMD3 performances versus the power sweep for the two-tone signal.

adjusted since the linearity at the 6 dB back-off region is closely related to the cancellation of the IM3 components generated by the carrier and peaking amplifiers. The peaking amplifier should be turned on at a 6 dB back-off level from the lower input power. To maintain the turn-on timing at the optimum back-off level, the base bias of the peaking amplifier needs to be increased because the threshold voltage of a transistor does not change much with the collector bias sweep.

The simulated optimum base currents of the carrier and peaking amplifiers to maintain the linearity at the high-power region with the $V_{cc}$ sweep are depicted in Fig. 5.31. As shown, the carrier current should be reduced linearly as the collector bias is reduced. However, the peaking current is increased with the reduced voltage. The adaptive circuits suitable to control the base current levels of the carrier and peaking amplifiers are shown in Fig. 5.32. The control voltage of the adaptive bias circuits is supplied from the output voltage of the multi-level DC-DC converter ($V_{\text{out.APT}}$). The adaptive bias circuits automatically generate the extracted optimum base bias points of the carrier and peaking amplifiers, respectively, as shown in Fig. 5.31. As explained, the optimum bias point of the carrier amplifier is decreased and that of the peaking amplifier is increased, as the collector bias is decreased.

## 5.6.2 Implementation and Measurement

The Doherty power amplifier is fabricated using an InGaP/GaAs HBT and the chip area is $1.25 \times 0.95\ \text{mm}^2$, as shown in Fig. 5.29B. Texas Instruments' LM3290 product is used for the multilevel DC-DC converter. Its output voltage ranges are from 0.4 to 3.81 V, and the highest efficiency is 95% with $V_{\text{out.APT}}$ of 3.81 V. Fig. 5.33 shows the measured

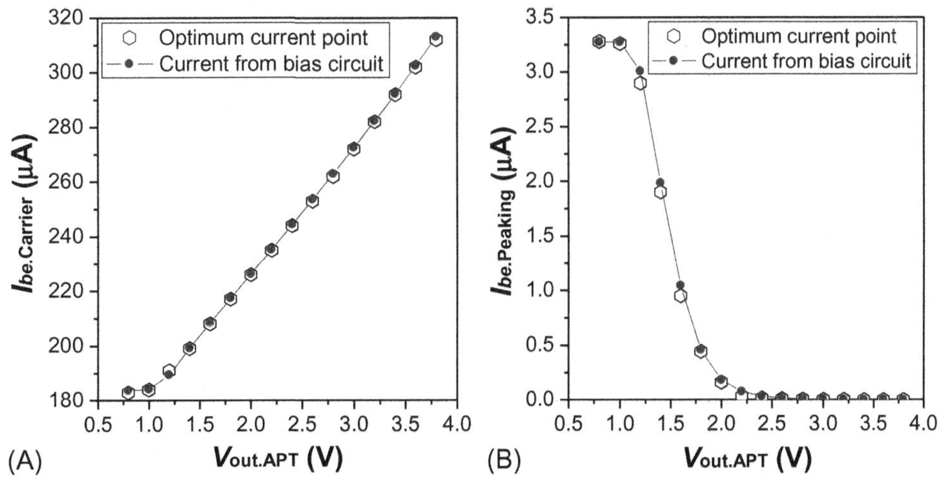


**Fig. 5.31** Simulated optimum bias currents and the currents generated by the adaptive base bias circuits for the (A) carrier and (B) peaking amplifiers versus the $V_{\text{out.APT}}$.

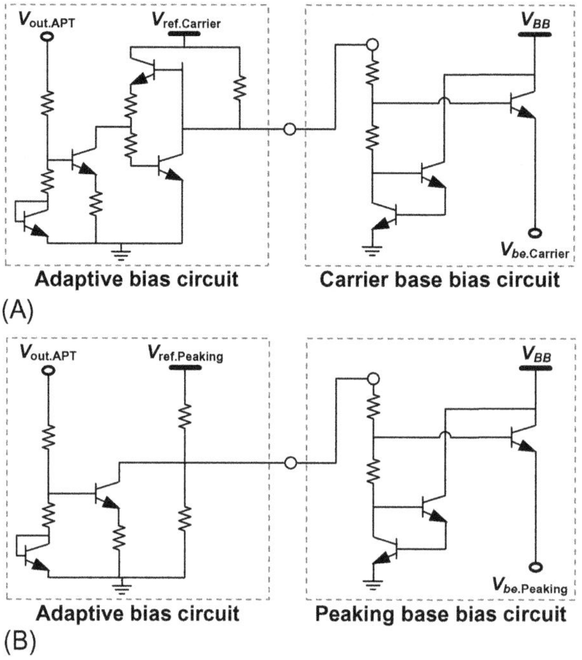


**Fig. 5.32** Schematics of the adaptive base bias circuits for the (A) carrier and (B) peaking amplifiers.

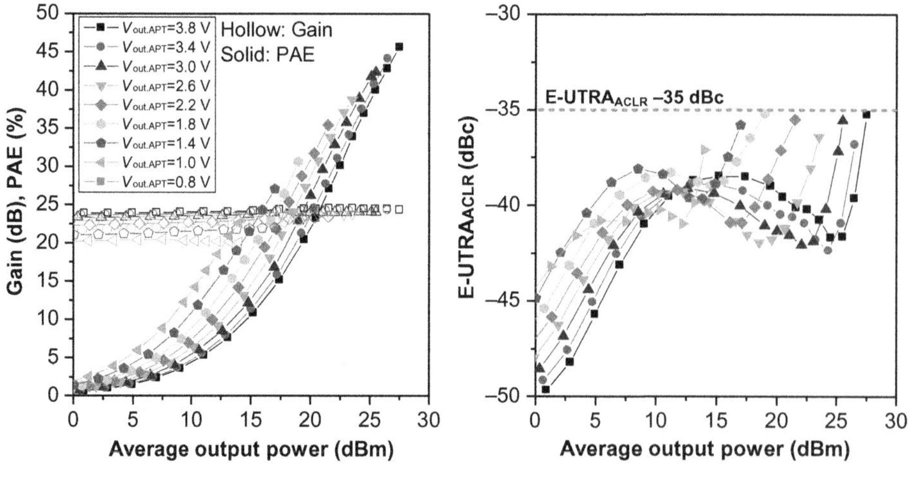


**Fig. 5.33** Measured results of the Doherty power amplifier with APT. The gain, PAE, and E-UTRA$_{ACLR}$ performances versus the power sweep for the LTE signal.

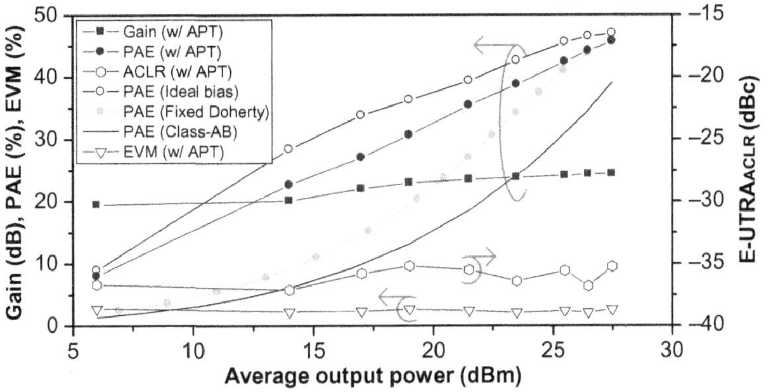


**Fig. 5.34** The performance of the Doherty power amplifier with APT, the comparison between the Doherty power amplifier with ideal collector bias, with fixed bias, and the class AB power amplifier.

performance of the Doherty power amplifier with various $V_{out.APT}$ at 1.9 GHz for a 10 MHz BW 16-quadrature amplitude modulation (QAM) 7.5 dB PAPR LTE signal. The power amplifier with $V_{out.APT}$ of 3.8 V delivers a gain of 24.4 dB, a PAE of 45.7%, an error vector magnitude (EVM) of 2.65%, and an evolved universal terrestrial radio access adjacent channel leakage ratio (E-UTRA$_{ACLR}$) of −35.3 dBc at an average output power of 27.5 dBm. The Doherty power amplifier with the APT shows a high efficiency at overall dynamic ranges for various $V_{out.APT}$ from 1.0 to 3.8 V. The E-UTRA$_{ACLR}$ specifications are satisfied for all $V_{out.APT}$ having more than −5 dBc margin. In comparison with the Doherty power amplifier with the ideal APT, the efficiency of the power amplifier with the APT is slightly lower due to the limited efficiency of the multilevel DC-DC converter as shown in Fig. 5.34. However, the efficiency at the whole back-off regions is significantly improved from the Doherty power amplifier with a fixed bias.

## FURTHER READING

[1] Y. Cho, et al., A handy dandy Doherty PA, IEEE Microw. Mag. 18 (September/October) (2017) 110–124.
[2] D. Kang, et al., Design of Doherty power amplifiers for handset applications, IEEE Trans. Microwave Theory Tech. 58 (8) (2010) 2134–2142.
[3] D. Kang, et al., Impact of nonlinear $C_{bc}$ on HBT Doherty power amplifiers, IEEE Trans. Microwave Theory Tech. 61 (9) (2013) 3298–3307.
[4] Y. Cho, et al., Linear Doherty power amplifier with an enhanced back-off efficiency mode for handset applications, IEEE Trans. Microwave Theory Tech. 62 (3) (2014) 567–578.
[5] Y. Cho, et al., Voltage-combined CMOS Doherty power amplifier based on transformer, IEEE Trans. Microwave Theory Tech. 64 (11) (2016) 3612–3622.
[6] Y. Cho, et al., Linear Doherty power amplifier with adaptive bias circuit for average power-tracking, IEEE MTT-S Int. Microw. Sympo. Dig., San Fansico, CA, 22–27, May, 2016.

[7] W. Kim, et al., Analysis of nonlinear behavior of power HBTs, IEEE Trans. Microwave Theory Tech. 50 (7) (2002) 1714–1722.
[8] D. Kang, et al., Broadband HBT Doherty power amplifiers for handset applications, IEEE Trans. Microwave Theory Tech. 58 (12) (2010) 4031–4039.
[9] D. Kang, et al., Design of bandwidth-enhanced Doherty power amplifiers for handset applications, IEEE Trans. Microwave Theory Tech. 59 (12) (2011) 3474–3483.

# INDEX

Note: Page numbers followed by *f* indicate figures.

## L



## N



## O



## P

## Q



## R



## S



## T



## V



## W

Printed in the United States
By Bookmasters